W9-BTE-191

CURRENT

OVERCOMPLICATED

OVER COMPLICATED

Technology at the Limits of Comprehension

Samuel Arbesman

CURRENT

CURRENT
An imprint of Penguin Random House LLC
375 Hudson Street
New York, New York 10014
penguin.com

ISBN 9781591847762

Printed in the United States of America
10 9 8 7 6 5 4 3 2 1

Set in Warnock Pro with Verb
Designed by Daniel Lagin

For Abigail and Nathan,
who will come of age in a world of wonders.

Never stop being excited by it.

Contents

OVERCOMPLICATED

INTRODUCTION

On July 8, 2015, as I was in the midst of working on this book, United Airlines suffered a computer problem and grounded its planes. That same day, the New York Stock Exchange halted trading when its system stopped working properly. *The Wall Street Journal*'s website went down. People went out of their minds. No one knew what was going on. Twitter was bedlam as people speculated about cyberattacks from such sources as China and Anonymous.

But these events do not seem to have been the result of a coordinated cyberattack. The culprit appears more likely to have been a lot of buggy software that no one fully grasped. As one security expert stated in response to that day's events, "These are incredibly complicated systems. There are lots and lots of failure modes that are not thoroughly understood." This is an understated way of saying that we simply have no idea of the huge number of ways that these incredibly complex technologies can go wrong.

Our technologies—from websites and trading systems to urban infrastructure, scientific models, and even the supply chains and logistics that power large businesses—have become hopelessly interconnected and overcomplicated, such that in many cases even those who build and maintain them on a daily basis can't fully understand them any longer.

In his book *The Ingenuity Gap,* professor Thomas Homer-Dixon describes a visit he made in 1977 to the particle accelerator in Strasbourg, France. When he asked one of the scientists affiliated with the facility if there was someone who understood the complexity of the entire machine, he was told that "no one understands this machine completely." Homer-Dixon recalls feeling discomfort at this answer, and so should we. Since then, particle accelerators, as well as pretty much everything else we build, have only increased in sophistication.

Technological complexity has been growing for a long time. Take the advent of the railroads, which required a network of tracks and a switching system to properly route trains across them. The railroads spurred the development of standardized time zones in the United States in order to coordinate the many new trains that were crisscrossing the continent. Before this technology and the complexity it entailed, time zones were less necessary.

But today's technological complexity has reached a tipping point. The arrival of the computer has introduced a certain amount of *radical novelty* to our situation, to use the term of the computer scientist Edsger Dijkstra. Computer hardware and software is much more complex than anything that came before it, with millions of lines of computer code in a single program and microchips that are engineered down to a microscopic scale. As computing has become embedded in everything from our automobiles and our telephones to our financial markets, technological complexity has eclipsed our ability to comprehend it.

In recent years, scientists have even begun to recognize the inextricable way that technology and nature have become intertwined. Geologists who study the Earth's rock layers are asking whether there is enough evidence to formally name our current time period the *Anthropocene,* the Epoch of Humanity. Formal title or not, the relationship between our human-made systems and the natural world means that each of our actions has even more unexpected ramifications than ever before, rippling not just to every corner of our infrastructure but to every corner of the planet, and sometimes even beyond. The totality of our technology and infrastructure is becoming the equivalent of an enormously complicated vascular system, both physical and digital, that pulls in the Earth's raw materials and emits roads, skyscrapers, large populations, and chemical effluent. Our technological realm has accelerated the metabolism of the Earth and done so in an extraordinarily complicated dance of materials, even changing the glow of the planet's surface.

We are of two minds about all this complexity. On the one hand, *we built* these incredibly complicated systems, and that's something to be proud of. They might not work as expected all the time, but they are phenomenally intricate edifices. On the other hand, almost everything we do in the technological realm seems to lead us away from elegance and understandability, and toward impenetrable complexity and unexpectedness.

We already see hints of the endpoint toward which we are hurtling: a world where nearly self-contained technological ecosystems operate outside of human knowledge and understanding. As a journal article in *Scientific Reports* in September 2013 put it, there is a complete "new machine ecology beyond human response time"—and this paper was talking only about the financial world. Stock market machines interact with one another in rich ways, essentially as algorithms trading among themselves, with humans on the sidelines.

This book argues that there are certain trends and forces that overcomplicate our technologies and make them incomprehensible, no matter what we do. These forces mean that we will have more and more days like July 8, 2015, when the systems we think of as reliable come crashing down in inexplicable glitches.

As a complexity scientist, I spend a lot of time being preoccupied with the rapidly increasing complexity of our world. I've noticed that when faced with such massive complexity, we tend to respond at one of two extremes: either with fear in the face of the unknown, or with a reverential and unquestioning approach to technology.

Fear is a natural response, given how often we are confronted with articles on such topics as the threat of killer machines, the dawn of superintelligent computers with powers far beyond our ken, or the question of whether we can program self-driving cars to avoid hitting jaywalkers. These are technologies so complex that even the experts don't completely understand them, and they also happen to be quite formidable. This combination often leads us to approach them with alarm and worry.

Even if we aren't afraid of our technological systems, many of us still maintain an attitude of wariness and distaste toward the algorithms and technologies that surround us, particularly when we are confronted with their phenomenal power. We see this in our responses to the inscrutable recommendations of an Amazon or a Netflix, or in our annoyance with autocorrect's foibles. Many of us even rail at the choices an application makes when it tells us the "best" route from one location to another. This phenomenon of "algorithm aversion" hints at a sentiment many of us share, which appears to be a lower-intensity version of technological fear.

On the other hand, some of us veer to the opposite extreme: an undue veneration of our technology. When something is so complicated that its behavior feels magical, we end up resorting to the terminology and solemnity of religion. When we delight at Google's brain and its anticipation of our needs and queries, when we delicately caress the newest Apple gadget, or when we visit a massive data center and it stirs something in the heart similar to stepping into a cathedral, we are tending toward this reverence.

However, neither of these responses—whether from experts or laypeople—is good or productive. One leaves us with a crippling fear and the other with a worshipful awe of systems that are far from meriting unquestioning wonder. Both prevent us from confronting our technological systems as they actually are. When we don't take their true measure, we run the risk of losing control of these systems, enduring unexpected and sometimes even devastating outcomes. Next time, the results of our failure to understand might not be as trivial as a frustrated *Wall Street Journal* reader being unable to access an article at the time of her choosing. The glitches could be in the power grid, in banking systems, or even in our medical technologies, and they will not go away on their own. We ignore them at our peril.

Technology, while omnipresent, is not pristine or unfathomable because of its creation by some perfect, infinite mind.

It is wonderfully messy and imperfect. And it is still approachable. We require a strategy to directly confront this situation.

My goal is to help each of us navigate a path between the two extremes of fear and awe, laying out an orientation toward our technologies that will allow us to make progress in how we approach them. It is an optimistic orientation, one that involves changing the way we think about these systems without falling into paralyzing fear or reverence.

This orientation will require us to meet our technologies halfway by cultivating a comfort with these systems despite never completely understanding them. This is the sort of humble comfort that dwells in ambiguity and imperfection, yet constantly strives to understand more, bit by bit. As we will see, this orientation involves, among other things, each of us thinking the way scientists do when examining the massive complexity of biology.

Despite all the overcomplication of the systems we vitally depend on, I'm ultimately hopeful that humanity can handle what we have built.

This book is why.

Chapter 1
WELCOME TO THE ENTANGLEMENT

On a winter day early in 1986, less than a month after the *Challenger* disaster, the famous physicist Richard Feynman spoke during a hearing of the commission investigating what went wrong. Tasked with determining what had caused the space shuttle *Challenger* to break apart soon after takeoff, and who was to blame, Feynman pulled no punches. He demonstrated how plunging an O-ring—a small piece of rubber used to seal the joints between segments of the shuttle's solid rocket boosters—into a glass of ice water would cause it to lose its resilience. This small piece of the spacecraft was sensitive to temperature changes, making it unable to provide a firm seal. These O-rings seem to have been responsible for the catastrophe that cost seven crew members their lives.

Contrast this example with another failure, one in the

automobile industry. In 2007, Jean Bookout was driving a 2005 Toyota Camry in a small town in Oklahoma when her car began accelerating uncontrollably. She attempted to brake, even using the emergency brake and leaving skid marks on the road. Her efforts did not stop the car, and it crashed into an embankment. Bookout was left terribly injured, and her friend and passenger, Barbara Schwarz, died.

Bookout's story is not a unique one. For a number of years, it seems that numerous vehicles manufactured by Toyota exhibited this strange and dangerous problem: they maintained or even increased speeds against the will and efforts of the driver. Multiple people died as a result of this "unintended acceleration." Several potential causes were proposed, including driver error, floor mats that could jam the gas pedal, and even a sticky gas pedal. But there were too many cases these causes couldn't account for: fewer than half of the affected models ever had any recalls for ill-fitting floor mats or sticky pedals, and there was no reason to believe drivers of Toyotas were so much more likely to err than drivers of other types of cars.

Toyota granted access to its proprietary and closely held software code to an embedded-software expert named Michael Barr. With the assistance of half a dozen other experienced engineers, Barr endeavored to explain what went wrong. The computer scientist Philip Koopman has also examined publicly available features of the design in order to understand what might have occurred. Both experts concluded that the massive complexity and poor design of Toyota's engine software was responsible for at least some of the unintended acceleration in these cars. No single piece or design could be pointed to as the clear-cut cause of this problem. Rather, there were several distinct problems that interacted, and the fault thus lay in the massive interconnectivity of the baroque structure of the computer code and the surrounding electromechanical systems in these cars. The complexity of this system made it difficult to understand the implications for these interacting pieces that, both individually and when combined, had deep issues and flaws. According to the evidence presented, Toyota could have been far more careful when building such a complex—in this case, unnecessarily complex—system.

This diagnosis couldn't feel more different from our familiar models of technical failure. Unlike the well-known story of Feynman's demonstration of the reason for the *Challenger* disaster, there was no single smoking gun that we could comfortably point to as the cause of the problem with Toyota's cars. Rather, a steady march of complicated components and failures of design, when combined, added up to a disaster for Toyota.

In fact, even when we can find a single cause for a failure, it actually may be somewhat of a red herring in today's complex systems. In 1996, an Ariane 5 rocket exploded, self-destructing thirty-nine seconds after launch. All four satellites on board were lost. Analysis of the failure revealed that the explosion was due to some older software code being used in the newer rocket under new conditions. But according to Dr. Homer-Dixon, no individual contractor was blamed. The explosion was less the fault of a single decision than of the incredible complexity of the entire system involved in launching this rocket into space. Other similar disasters, such as the Three Mile Island nuclear disaster, might also have an identifiable cause, but when it comes down to the real reason for the failure, it's more accurate to say it was the system's massive complexity, rather than any single component or choice.

When we think of untangling massive complexity, we are drawn to the popular narrative of the *Challenger.* Even though the launch and operation of a space shuttle was an incredibly complex matter, we feel that by applying our ability to scrutinize sophisticated systems, break them down, and determine how they work and sometimes fail, we should be able to understand them. We owe this overconfidence to an idea that many of us are ensorcelled by: the unlimited potential of the human mind. We believe that if we just work hard enough, we can achieve a perfect understanding of everything around us, especially what we ourselves have built.

We see this sentiment in what is termed the Whiggish view of progress, described by the writer Philip Ball as the belief that humanity is on "a triumphant voyage out of the dark ages of ignorance and superstition into the light of reason." This view is found within science as well as among technophiles, as the historian Ian Beacock writes: "The tech industry tells a

Whiggish tale about the digital ascent of humanity: from our benighted times, we'll emerge into a brighter future, a happier and more open society in which everything has been measured and engineered into a state of perfect efficiency." Surely, say those who adhere to such a viewpoint, this growth in efficiency and productivity presumes our continued ability to understand the phenomenal engineering we have constructed for such an uplifting purpose. This Whiggish perspective is related to the modern mind-set described by the sociologist Max Weber: a sense that "the world is disenchanted," that "one can, in principle, master all things by calculation."

But more and more, this way of approaching complexity just doesn't work. We are in a new era, one in which we are building systems that can't be grasped in their totality or held in the mind of a single person; they are simply too complex. We are finding ourselves, expert or not, more often in the realm of Toyota's "unintended acceleration," where this old way of thinking will no longer suffice. These situations are not at the edges of our experience; they suffuse our lives.

What Complex Systems Are—and Aren't

It's worth taking a moment to talk about what we mean by "complicated" and "complex" systems. While in this book (and even in its title) I use these terms—complex and complicated—more or less interchangeably in their colloquial sense, there are important distinctions.

Imagine water buoys, tied together, floating in the water. As a boat goes by, its wake generates small waves that begin moving one buoy, then another. But each buoy does not act alone. Its own motion, since it is connected by rope to other buoys of different weights and sizes, causes changes in the undulations of the others. These movements can even cause unexpected feedback, so that one buoy's motion eventually comes back, indirectly, to itself. The boat's simple wake has generated a large cascade of activity across this complex network of buoys. And if the boat had sped by in just a slightly different way—at another speed or angle—the motions of the buoys might have been entirely different.

Now, let's say we pull up the buoys and toss them onto the dock. Their arrangement is precise and potentially hard to describe—it might require many paragraphs to write out the arrangement of these buoys, complete with diagrams, so that someone could re-create this specific configuration—but there's nothing interesting going on here. There are no cascading effects, no feedback, no process happening within this sophisticated network. It's just a bunch of stuff that can float, sitting in a pile on a dock.

The buoys in the water form a *complex* system. The buoys on the dock? Their arrangement is simply *complicated.* For a system to be complex, it's not sufficient for it to contain lots of parts. The parts themselves need to be connected and interacting together in a tumultuous dance. When this happens, you see certain characteristic behaviors, hallmarks of a complex system: small changes cascade through this network, feedback occurs in the complex system, and there is even a sensitive dependence on the initial state of this system. These properties, among others, take a system from complication to complexity.

Here's another way to think about this distinction: living creatures are complex, while dead things are complicated. A dead organism is certainly intricate, but there is nothing happening inside it: the networks of biology—the circulatory system, metabolic networks, the mass of firing neurons, and more—are all quiet. However, a living thing is a riot of motion and interaction, enormously sophisticated, with small changes cascading throughout the organism's body, generating a whole host of behaviors. Furthermore, even if a system is dynamic—such as a bunch of unconnected buoys floating in the water—if there is no interconnection, potential for feedback, or other such properties, we are still in the realm of complication, not complexity.

If you define technology as any sort of system that humans have built and engineered for a specific purpose, you notice that almost all of today's most advanced technologies are complex systems: dynamic, functionally intricate, of vast size, and with an almost organic level of complexity. These complex systems are all around us, from the software in our cars to the computers found in our appliances to the infrastructure of our cities. We have software projects that are massive, highly interconnected, and could fill encyclopedias—Microsoft Office alone has been estimated to be in the tens of millions of lines of computer code. The road system in the United States has 300,000 intersections with traffic signals and is the substrate for the constantly churning turmoil of transportation that spans a continent. Autocorrect, which we often deride as being hopelessly stupid for its failures, is actually incredibly advanced, relying on petabytes of data (a petabyte is a million gigabytes) and complex probability models. Our legal constructions have also become more complex over time, with the number of pages in the federal tax code numbering over 74,000 as of 2014. This vast legal network is profoundly complex, with numerous interconnections and a cascade of interacting effects on taxpayers, the functionality of which no person could completely understand in its entirety.

The intricacy in the complex technological systems that suffuse our lives is often a good thing. Within their vast complexity we find resilience and sophistication. These systems often possess many features and fail-safes that help them deal with anything that comes their way. These systems also provide us a life that the royalty of the ancients couldn't imagine. They allow us to automate drudgery, bring water and power to our homes, live in perfect climates year-round, and summon information instantly.

But what does it mean to *understand* these complex systems? Understanding a phenomenon or system is not a binary condition. It exists across a rich spectrum. For example, you can understand a system as a whole, in its broad strokes, but not the details of the parts within it; you can understand all of its parts but not how they function together as a whole; you can understand how the parts are connected, or perhaps only the effect of these connections. Further, each of these components of understanding involves specific activities: *describing* how something works, *predicting* its future actions to varying degrees, and *replicating* it through a model given enough time and resources.

To return to our example, you might only understand the behavior of two or three buoys connected together, or even the motion of a single buoy in great detail, rather than the vast network of all the combined buoys. You might be able to describe the motion of the buoys without predicting their behavior. In software, you might understand a few modules in a given program really well—the one that calculates the value of pi, or the one that can efficiently sort a list of numbers—but not necessarily how they all work together. Often, we can grasp only some of these components of understanding, rather than all of them.

In addition, understanding is not static; it can improve

with training. Someone who has never played chess before will look at most chessboard configurations and be unable to distinguish the differences between a hopeless muddle, the endgame, and threats to your king. A novice or intermediate player, however, will begin to see patterns in the pieces, and the current thread of the game. A master will see whole patterns at once, and multiple future patterns that could evolve from them, surveying the game and identifying potential moves and weaknesses. With sufficient training, one's view of a chessboard goes from an array of pieces to a configuration that has white to checkmate in three moves. Training and expertise can actually change how we see the world and how we understand it.

We see the same situation with the systems we've built. Pages of computer code can be either gobbledygook or a beautiful solution to a difficult problem, depending on what you know. But when we fail to have a complete understanding, we fall short in a specific way: we encounter unexpected outcomes.

Take the Traffic Alert and Collision Avoidance System (TCAS), which was developed to prevent airplanes from crashing into each other in the sky. TCAS alerts pilots to potential hazards, and tells them how to respond by using a series of rules. In fact, this set of rules—developed over decades—is so complex that perhaps only a handful of individuals alive even understand it anymore. When a TCAS is developed for a new airplane, a simulation is used to test its effectiveness. If the new system responds as expected after a number of test cases, it receives a seal of approval and goes into use.

While the problem of avoiding collisions is itself a complex challenge, the system we've built to handle this problem has essentially become too complicated for us to understand, with even experts sometimes reacting with surprise to how it responds in certain situations.

When an outcome is unexpected, it means that we don't have the level of understanding necessary to see how it occurred. If it's a bug in a video game, this can be delightful or even entertaining. But when we encounter unexpected situations in the complex systems that allow our society to function—the infrastructure that provides our power and water, or the software that allows financial transactions to occur, or the program that prevents planes from colliding midair—it's not entertaining at all. Lack of understanding becomes a matter of life and death.

While there are natural variations in our abilities to understand the world, with geniuses capable of incredible intuitive leaps that the rest of us struggle to grasp, we, as humans, still have cognitive limits. Increasingly, as we build technological systems that are ever more complicated and interconnected, we become less able to understand them, no matter how smart we are or how prodigious our memory, because these systems are constructed differently from the way we think. Humans are ill-equipped to handle millions of components, all interacting in huge numbers of ways, and to hold all the implications in our heads. We get overwhelmed, and we fail.

The Entanglement

In our era of modern machines, the non–technologically savvy among us occasionally resort to superstition and wishful thinking in an attempt to understand technology. For instance, there is invariably one person in a family who is blamed for a computer not working. Sometimes their touch mucks things up; sometimes even their mere presence is deemed to have caused technology not to function as it should. A child comes home from college, and the printer stops working. Or a parent visits and the mouse ceases to function.

Then there's the opposite issue: when a problem inexplicably vanishes the minute a solution is at hand. You bring a malfunctioning machine to technical support, and as soon as they touch it, the problem is nowhere to be found. But when you bring it home, you discover you still have a broken device.

This is the experience of the layperson; in our absence of technical expertise, the inner workings of these machines can appear somewhat magical. If we, as users of these systems, don't know all the inner complications, it doesn't matter. When failures happen, we can half-seriously assume that someone is having a perversely demonic effect on our machines. And even if we don't make this assumption, we are comfortable recognizing that at least the expert knows what's going on, and can reduce the mysterious poltergeist to a case of misfiring motherboards.

Unfortunately, this attitude is no longer reserved for the common person; it occurs even among the developers of technology themselves. The engineer Lee Felsenstein has told the story of an engineering manager who had to be removed from the room whenever a piece of software was being demonstrated, because his presence caused things to malfunction. The designers simply had no idea why this manager's presence made things go bad. As Felsenstein noted, this type of computationally unexplainable failure "falls into the area of metaphysics." These engineers simply didn't know what was going on, and felt compelled to wave their hands in the general direction of philosophical musings on the nature of being.

They are not alone. The computer scientist Gerard Holzmann has much the same feeling:

> Large, complex code almost always contains ominous fragments of "dark code." Nobody fully understands this code, and it has no discernable purpose; however,

> it's somehow needed for the application to function as intended. You don't want to touch it, so you tend to work around it.
>
> The reverse of dark code also exists. An application can have functionality that's hard to trace back to actual code: the application somehow can do things nobody programmed it to do.

Similarly, in the legal field, we are currently in a situation where, according to the lawyer and author Philip K. Howard, "Modern law is too dense to be knowable." But we don't even need to examine the frontiers of our technologies to find such examples. As the writer Quinn Norton has noted, even your average desktop machine is "so complex that no one person on Earth really knows what all of it is doing, or how."

In recent decades, the inexplicable, not just the complicated, is turning up more and more in the world of our own creations, even for those who have built these systems. Langdon Winner notes in his book *Autonomous Technology* that H. G. Wells came to believe late in life "that the human mind is no longer capable of dealing with the environment it has created." Wells concluded this in 1945, discussing primarily human organizations and societies. This problem has become even more acute in recent years, through the development of computational technologies to a level that even Wells might have had difficulty imagining.

The computer scientist Danny Hillis argues that we have moved from the Enlightenment to the Entanglement, at least when it comes to our technology: "Our technology has gotten so complex that we no longer can understand it or fully control it. We have entered the Age of Entanglement. . . . Each expert knows a piece of the puzzle, but the big picture is too big to comprehend." Not even the experts who have actually built them fully understand these technologies any longer.

The Limits of Abstraction

When we build complex technologies, one of the most powerful techniques for constructing systems is what is known as *abstraction.* Abstraction is essentially the process of hiding unnecessary details of some part of a system while still retaining the ability to interact with it in a productive way. When I write a computer program, I don't have to write it in machine code—the language written out in binary code that each specific computer uses for its instructions. Instead, I can use a programming language such as C, one that can be more easily read by people and yet still be translated into machine code. In many cases, I don't even need to know what specific machine my program might run on: these details have all been taken care of by other programs that interact at deeper levels with the machine. In other words, these details have been abstracted away.

This kind of abstraction occurs everywhere in technology, whether we are interacting with a user-friendly website whose innards we don't care about, or plugging a toaster into an outlet anywhere in the country and receiving electrical current. I don't need to know where the electricity came from or where it was generated, just as I don't need to know the specifics of how a search engine generated its results. As long as the interface is a logical and accessible one, I can focus on the details of whatever I am building (or fixing) and not worry about the complexity that lies beneath. Abstraction allows someone to build one technology on top of another, using what someone else has created without having to dwell on its internal details. If you are a financial analyst using a statistical package to examine datasets or an app developer using pre-written code to generate fancy graphics, you are using abstraction.

Abstraction can bring us the benefits of specialization. Even if a system has millions of interacting components, those working to build or maintain it don't necessarily need to know how it all works; abstraction allows them the luxury of needing to know only about the specific part that they are focused on. The rest of the details are, again, abstracted away.

Unfortunately, in the Entanglement, abstraction can—and increasingly will—break down. Portions of systems that were intended to be shielded from each other increasingly collide in unexpected ways.

One of the places where we can clearly see this is in the

financial realm. Today's markets involve not just humans, but large numbers of computer programs trading on a wide variety of information at rates faster than what people could do manually. These programs interlock in complicated ways, making decisions that can cascade through vast trading networks. But how are the decisions made on how to trade? By pouring huge amounts of data into still other programs, ones that fit vast numbers of parameters in an effort to squeeze meaning from incredible complexity.

The result can be extreme. Take the so-called Flash Crash, when, on May 6, 2010, the global financial market experienced a massive but extremely rapid fluctuation in the stock market, as large numbers of companies lost huge amounts of value, only to regain them instants later. This crash seems to have involved a series of algorithms and their specific rules for trading all interacting in unexpected ways, causing a trillion dollars in lost value for a short period of time. Complex though they are, these systems do not exist in a vacuum. In addition to being part of a larger ecosystem of technology that determines when each specific equity or commodity should be traded, our financial systems are also regulated by a large set of laws and rules. And of course, this collection of regulations and laws is itself also a system—a sophisticated and complex one—with massive numbers of laws that are interdependent and reference one another in precise and sometimes inscrutable ways.

Furthermore, the infrastructure that allows these trades to

be executed is built upon technologies that have grown over the decades, yielding a combination of the old and the new: traders at hoary physical stock exchanges coexist with fiber optic cables in this system. When we are building computer programs that trade at high speeds, understanding how to do this effectively doesn't just require knowledge of computer science, complex financial instruments, and laws and regulations, but now also a deep understanding of physics, because the speed of light in different materials plays a role at these trading speeds. As a result, there is no single person on the planet who fully understands all the interconnected systems of the financial world, and few completely understand even a single one.

For many situations, each person working with a system needs to understand only a subset of the system well, or even some of it at just a superficial level. Some programmers at a financial firm might know only how to work on the system that's actually doing the trading, while having almost no need to understand the physical infrastructure of the computers in the company. Others might focus on specific pieces of software that allow messages to pass from outside their firm to the algorithms that operate internally, and have only a passing familiarity with most everything else. And a lawyer working for the company might need to know about the laws that regulate certain types of trading, but would have no need to know the details of the software, servers, or fiber optics. Abstraction serves us nicely.

Understanding something in a "good enough" way can be just fine—most of the time. But as we build systems that are more and more complicated, the different levels at which the systems and their subsystems operate increasingly interact in some sort of multidisciplinary mess. In particular, as things become more interconnected, it becomes difficult to know whether a cursory or incomplete understanding is really sufficient. In the Entanglement, things collide across the many levels of abstraction, interacting in ways we can't imagine. This web of interactions produces what's known in complexity science as *emergence,* where the interactions at one level end up creating unanticipated phenomena at another. This is common in complex systems of all types, such as, for instance, when insects move together to create the emergent behavior of a swarm. It is also particularly clear in our financial systems, which involve multiple factors down to the level of the speed of transmission across individual wires and up to the level of planet-wide computational interactions. It's too complicated to really know whether being able to abstract away the details will be sufficient.

When those tiny little details deep within the system rise up like miniature demiurges and ruin some other portion of the technological system that we have constructed, we can no longer rely on understanding only part of the system. Hierarchies and abstraction, which have helped us manage complexity, now are increasingly collapsing in the face of the messy interaction of the Entanglement.

So is there any hope in sight, any way of returning from this muddle? Or are we doomed to contemplate these proliferating systems with a profound and ineffable horror?

Most of us think it's okay if we don't fully understand these technological systems, if we don't know the details of our urban infrastructure or how the hardware of the iPhone registers our finger's touch, or even how the morass of laws and regulations allows international commerce to occur. The mechanics of complicated systems, we assume, are unimportant as long as we are able to use them. But it's one thing to not understand how your new gadget works; it's another thing entirely when *no one* truly understands that gadget. While many of us continue to convince ourselves that experts can save us from this massive complexity—that they have the understanding that we lack—that moment has passed.

Our old patterns of making sense of these systems—the *Challenger*-style modes of thinking—are now hopelessly inadequate. The Entanglement is not off on the remote horizon, something to be rarely encountered. It is here, all around us. Each of us is going to need new ways of thinking about these technologies, even the ones for which we have blithely outsourced understanding to experts.

Although the dawn of the Entanglement looks pretty grim,

I'm hopeful: we can learn to handle these systems, at least to some degree.

But to really understand this era we have created for ourselves, we need to take a step back and identify the forces that are propelling us deeper into complexity—and preventing us from comprehending it.

Chapter 2
THE ORIGINS OF THE KLUGE

In order to use the Internet, we must endure—at least indirectly—what can only be described as a mess. What was to become the Internet first began to be developed during the 1960s. It had an ingenious design, allowing it to be decentralized and to easily pass packets of information between different machines. This allowed smaller networks to be interconnected, with numerous protocols developed to enable this to happen efficiently.

Today, we no longer use the Internet the way it was originally developed. Take security, for example. A system developed by researchers to communicate is not ideal for a high volume of smooth and secure commercial transactions. To compensate for the Internet's flaws, we have developed different mechanisms on top of its basic infrastructure to allow these commercial transactions to occur, including a whole slew of ways of encrypting and decrypting information that must remain private, as well as ways of transferring money in the digital realm. Happily, the system does work. But beneath the user interfaces of a website lurks a bizarre and complicated structure. Sometimes we even see this mess directly, as users, when we see warnings about security certificates. Things work, but they are far from pretty.

Similarly, the language used to construct our websites—HTML—was never designed to handle slick interactive web-based applications like Google Docs. These applications have been constructed, but at a cost: we have had to weld a fantastically baroque edifice atop a simple system. To get a glimpse of this underlying complexity, one need only examine the source HTML of Google's homepage. While this webpage looks clean and elegant when viewed by a browser, there's a huge amount going on below the surface. Last time I checked, the code for Google.com is well over 100,000 characters long and would take more than fifty pages to print in its entirety.

Even email, which seems relatively simple, has evolved far beyond its decades-old roots, with features such as message threads grafted on top of its original structure. As *Slate* interactives editor Chris Kirk noted after embarking on an ill-advised attempt to build his own email client program: "Though innovation in email is happening, it's characterized by features balancing cunningly and sometimes haphazardly atop an antiquated system—features that attempt to either restore email to its original metaphor or evolve it into something else entirely."

Computer science and engineering have a term, *kluge,* for a cobbled-together, inelegant, and sometimes needlessly complicated solution to a problem. A kluge works, but it isn't pretty. Something may have been elegantly designed in its first iteration, but changes over time have complicated its structure, turning it into a Rube Goldberg–style jury-rigged mess.

There are kluges in every realm of technology, far beyond the structure of the Internet, from transportation to medical devices. Or even the wiring of your home entertainment system, which might work, but requires several remote controls and a Gordian knot of wires that is preferably located out of sight.

Consider the kluges of the American legal code, a technological system that has been built for a specific purpose but is far from elegant. Just as computer code is the written description of the operation of a piece of software, laws and regulations are technologies—essentially written embodiments of code as well.

The United States Constitution is a wonderfully elegant document. In only a handful of pages it lays out the foundation for a representative democracy. However, the Constitution is not the end of the story. The collection of federal laws that guide our country, known as the United States Code, is what has grown within this framework. This collection of laws has developed over the years to elaborate on the general principles of the Constitution as well as to handle specific situations. For example, while the Constitution states that Congress has the power to establish the postal service in the span of a single phrase, the United States Code spends more than 500 pages on the section that details this part of the government. This body of law includes everything from the bureaucratic details of the Postal Service to the specifics of postage. As a whole, the United States Code is far more complicated than America's founding document. It has been increasing in both size and interconnectivity, and is now more than 22 million words long, with more than 80,000 connections between one section and another.

We can see massive growth in complexity over time nearly everywhere we look. And in general, as a complex system becomes big enough, it ends up becoming a kluge of one sort or another, as we shall see. The airplane the Wright brothers built in 1903 was a masterpiece of simplicity, constructed with a small number of parts and weighing only 750 pounds including the pilot. A Boeing 747-400 has 147,000 pounds of aluminum, 6 million individual parts, and 171 miles of wiring. More generally, during the past 200 years the numbers of individual parts in our most complicated manufactured machines have increased massively.

What about software, which undergirds the modern technological systems of every aspect of our lives? One common way to measure the complexity of software is through the number of lines of code it takes to write a program. According to some estimates, the source code for the Windows operating system became ten times longer over the course of about a decade. The image-editing software application Photoshop has exploded in size over the course of twenty years, growing to nearly forty times the number of lines of code it had in 1990.

If we look at the telephone network, this kind of growth happened rapidly there as well, with huge levels of complexity appearing quickly. By the 1920s, the American telephone system already had about 3 million miles of toll circuits and about 17 million telephones. The telephone had only been invented several decades earlier, and already its technological ecosystem was enveloping the country.

Each of these systems is an engineered technology, crafted by generations of experts for a specific function. One might assume that if these systems are being designed rationally, they should be logical, elegant, and even simple; they'd be more predictable, and easier to fix. And yet, despite our best efforts, our technology becomes ever more complex and complicated. This is not happening by accident. There are several forces intrinsic to technological development that propel us ever deeper into complexity. This tendency is by no means a law of physics, like gravity, but the forces that make our systems more complicated over time are so strong, often overcoming our desires for something less complex, that they almost do feel like inexorable physical laws. But why is this so?

In this chapter I explore several forces that make systems more complicated over time. On the surface these forces seem eminently reasonable. Each individual change brought about by these forces might allow a technological system to adapt to changing circumstances, continue to operate in a new environment, or provide additional usefulness. But ultimately these forces replace our once-elegant solutions with messy kluges. No matter how hard we try to avoid this outcome, we are going to end up with increasing complexity in the technologies that touch every aspect of our lives.

To begin with, the most obvious reasons we end up with increasingly complex systems over time are the twin forces of *accretion* and *interaction:* adding more parts to something over time, and adding more connections between those parts.

Accretion

In the years leading up to January 1, 2000, many engineers were concerned with fixing the Y2K bug. In a nutshell, when a piece of software that used only two digits to store the year—rather than four—rolled over into the year 2000, it would assume it to be 1900 instead, and potential problems could arise. No one was more focused on handling this problem than those at the Federal Aviation Administration. If the air traffic control system began to malfunction come the New Year, that would be a huge problem. So the FAA began to examine its computers, testing date changes to see what would happen when the system thought it was 1900.

During the testing process, they discovered that one particular type of machine in their systems—the IBM 3083—was particularly tricky to fix. Among the issues, according to the head of the union representing the FAA technicians: "There's only two folks at IBM who know the micro-code, and they're both retired." That's because the IBM 3083 was a mainframe machine installed in the 1980s, with software dating from years earlier. In other words, as of the late 1990s, the systems that were responsible for properly routing our airplanes were using computers with code that almost nobody was familiar with any longer.

This is hardly a surprising story. Over and over, large systems are built on top of somewhat smaller and older systems. As long as the pieces work reasonably well, little thought is given to the layering of new upon old, the accumulation and addition of piece after piece. According to one source, as of 2007 the machines at the Internal Revenue Service responsible for processing tax returns used a computer repository system developed during the Kennedy administration of the early

1960s. A separate IRS system was built in the 1970s and last overhauled in 1985. Similarly, the final space shuttle mission was supported by five IBM machines whose computational power pales by comparison with today's average smartphone. And yet we continue using these kinds of software and technology.

In *The Mythical Man-Month,* originally published in 1975, Frederick P. Brooks Jr. examined software design and the management of programming projects. In the book he quotes the maxim "Add little to little and there will be a big pile." Each individual design decision, whether accounting for an exception or providing a new feature, may seem like a small and separate choice. Each one makes sense: it fixes a problem or creates some new and exciting functionality for users. But as these choices accumulate, they add up, and you eventually get a large pile. This happens no matter what type of large technological system we look at, from transportation to energy to agriculture.

A large pile of rocks, for example, is not necessarily a problem. It can be unwieldy and messy, but it doesn't have to be hard to understand. The problem is when the large pile we've created behaves in unexpected ways, such as when we get an avalanche. Unfortunately, this is often what happens when we add successive pieces to a technology. It grows not only bigger but less predictable as it accretes.

I first remember seeing the word "accretion" in relation to

how a planetary system is formed, condensing from a spinning mass of dust and gas. This process of accretion, of adding bits and pieces—sometimes very old—doesn't just create planets; technological growth is also a process of accretion.

One result of accretion is what is known as *legacy code* or *legacy systems:* outdated machines and pieces of technology that are still with us long after they were first developed, like those at the IRS. These antiquated systems are not exceptional or rare. They have been cobbled together and accreted over many years and are found everywhere we look, from the programs responsible for scientific simulations to parts of our urban infrastructure. For example, our cities can have water mains that are over one hundred years old coexisting alongside newer ones. In the case of computers, technological systems often rely on machinery that is no longer manufactured and code written in programming languages that have long since been retired. Many pieces of scientific software exist as legacy tools, often written in Fortran, a powerful but archaic programming language. Given the speed with which technology moves, reading Fortran is almost the computational equivalent of being well-versed in Middle English.

To quote the *Whole Earth Catalog* creator Stewart Brand in *The Clock of the Long Now:* "Typically, outdated legacy systems make themselves so essential over the years that no one can contemplate the prolonged trauma of replacing them, and they cannot be fixed completely because the problems are too complexly embedded and there is no one left who understands the whole system."

When we are left with a slowly growing, glitch-ridden legacy system, we can only gingerly poke it into doing our bidding, because those who designed it are long gone. It is so completely embedded into other systems that removing it appears far worse than living with its quirks. These systems, when unwieldy enough, are sometimes even referred to as *crawling horrors,* in deference to the unspeakable monsters from the stories of H. P. Lovecraft.

We see a parallel to legacy code in the law. In our legal systems, laws are modified or amended over time, tweaked for changing circumstances, leaving laws from decades earlier that are still relevant. Regulations around traffic on the Internet are derived from a law passed in 1934. As laws accrete over time, a legal system becomes a kluge—it gets the job done, but it is far from elegant.

In fact, the tax code is so complex that the law has recognized this fact. The number of pages of instructions for the 1040 tax form has exploded, from two in 1940 to more than 200 in 2013. If you make an error in your taxes in good faith simply because the rules and provisions are so complicated, the Supreme Court has ruled that you cannot be convicted for willful failure to file tax returns. Essentially, it is more efficient for the law to make these klugey patches on the overcomplicated tax code than to overhaul it entirely from scratch to make it more user-friendly.

Or consider the overall growth in regulations enacted by various departments and agencies of the government, such as the Environmental Protection Agency. For example, if you look only at the number of pages in the *Code of Federal Regulations*—the collection of rules from these many agencies—this number has gone from fewer than 25,000 to more than 165,000 in the past fifty years.

We see a similar growth in bureaucracies and administration. In an article from the 1950s, *The Economist* proposed something known as Parkinson's Law, which is a quantitative description of the multiplication of bureaucrats over time. While the article is somewhat tongue-in-cheek, it is buttressed with data, and concludes that governmental bureaucracies are essentially compelled to grow at the rate of about 5–6 percent more people per year. Managing an ever-growing bureaucratic structure seems to only increase in complexity.

In fact, those in the software world have enshrined this idea of accretion and accumulation to the level of a general rule. When it comes to software system evolution, these systems grow over time, unless there is an active attempt to simplify them.

So why don't we periodically sweep away all the complexity

and start from scratch? Some reasons are practical—there is simply not enough time to rewrite a piece of software, for example, instead of just releasing a patch on whatever came before it. I highly doubt that anyone at Microsoft would be fine with rebuilding Word from scratch if it would take many years. When there are constraints and trade-offs in time, effort, and money, a system often ends up being modified in a way that is "good enough," which generally means that things keep on getting tweaked and modified over time, just as lawmakers have done with the U.S. tax code. You build on top of what came before: our cities have gas pipes that are over a century old, transit systems run on 1930s technologies, and abandoned subway stations lurk beneath large cities.

But sometimes it is simply too difficult, or even dangerous, to start from scratch; no one understands the essential contributions of all the older pieces that a system relies on, and so to design an untried new system from the beginning would be foolish and even debilitating. For example, imagine a sophisticated banking software system that was designed many decades ago and has slowly been adapted to advancing technologies, from new computers and operating systems to the ubiquitous presence of the Internet. While the underlying core of the system was never intended for our modern era, these pieces are now far too embedded to remove. And so we must accept the general principle that systems become more complicated over time, whatever the technological system we look at.

But when we look at legacy code in our technologies—whether in software or in a body of laws—the real complexity isn't simply the result of growth in size over time. Accretion needs to incorporate another factor that complicates our technological systems: interaction.

Interaction

Students are currently inundated by exhortations to learn how to code. Computer programming is the future, educators and technologists tell us. It is used in the operation of everything around us, from our cars to our ovens. And in certain circumscribed ways, the experts are correct. Programming can teach you a structured way of thinking, and it can also provide a guide to what is truly possible with technology. If I say there's a new app that can do X, if you've coded, even a little bit, you're that much closer to knowing if my claim is reasonable or utter nonsense.

If you've coded before, you've also glimpsed how computer programs resist our efforts at simplification. As computer programs become larger and more sophisticated, they become more complicated, with more parts that all come together in ways that can be hard to grasp. We have a wide variety of techniques and technologies to make sure that these large programs are still manageable—things such as version control, bug-tracking software, or tools for communication across a team—but these are often rearguard efforts in a losing battle. For not only does software code accrete, but each of these components interacts with all the others. Far from each new addition existing in a pristine vacuum, a computer program is a massively interconnected system: it interacts with itself and with other pieces of software. We add layer upon layer to older code, using it in new and unexpected ways, stitching the layers all together.

Sometimes this process of interaction (and unexpected interaction) is exacerbated by the programming languages themselves. One problem even beginning programmers are familiar with is that of the command GOTO. In the programming language BASIC, a program can easily jump from one line to another using GOTO. When a GOTO is included in the code, the program's flow can easily be redirected from line to line, leaping from one point in the code to another. Making this happen is simple and easy, and has even been described as a "godsend" for those from the liberal arts who are testing their abilities in computational thinking.

In a small program, using GOTO to jump around is fine. But as a program becomes bigger, using GOTO ends up tying the code into complicated knots that are very hard for even a skilled programmer to untangle. You end up with what is known as *spaghetti code* because it is looped together, difficult to unravel, and hard to understand completely. It becomes nearly impossible to figure out the order of the instructions that will be executed in the computer program, leading to unexpected and incomprehensible behavior.

A simplifying construction, while it works in isolation and for small situations, has a way of escaping its purposes. A system or component gets used more often and in different ways than we could have possibly anticipated, leading to a far more complicated situation than initially envisioned. GOTO goes from being a wonderful shortcut to something considered not only inelegant, but actually "harmful."

There are many methods designed to help make sense of and impose order on the systems that we build, including the use of more sophisticated computer languages. Spaghetti code in most professionally constructed systems should be a thing of the past. However, interaction continues to occur because of the ease of interconnection and the accretion of the various pieces. The dynamics of a highly interconnected system—how information flows through it and what parts will interact with others—become incredibly complicated and unpredictable. To return to the software inside Toyota vehicles, according to one metric, numerous pieces of code within these vehicles are simply beyond the possibility of testing, because of this interconnection and interaction.

We see this same process of increased interconnection in other types of technological systems that we build. In our legal system, because each new law or regulation interacts with those that have come before it, we find ourselves in situations where it becomes difficult to predict the effects of individual laws.

The lawyer Philip K. Howard has examined the story of the Bayonne Bridge, which crosses a channel between New York and New Jersey. This nearly hundred-year-old bridge is too low to allow modern container ships to pass beneath it in order to reach the port of Newark, a critical commercial hub. So what can be done? Among the proposed solutions to this problem was one that involved retrofitting, essentially raising the bridge to the necessary height. It was the cheapest solution and the one that won out, back in 2009. But construction did not commence for several years, because of a combination of accretion and interaction. The rules and regulations dictating the procedure for the bridge's renovation involved forty-seven permits from nineteen different governmental entities—everything from environmental impact statements to historical surveys. Similar cases occur elsewhere, with some public projects taking around ten years to be approved because of the number of rules and processes that bear on such situations. This is particularly concerning when, as Howard notes, replacing decaying infrastructure can save lives.

Michael Mandel and Diana Carew of the Progressive Policy Institute in Washington, DC, have referred to this growth of rule systems as "regulatory accumulation," wherein we keep adding more and more rules and regulations over time. Each law or rule individually might make sense, but taken together they can be debilitating because of their interaction, and can even collide in surprising and unexpected ways.

We increasingly interconnect not only the components of a single piece of technology, but also different pieces of software and technology—a higher-order type of interconnection. This is the concept of *interoperability*.

It is often a good thing to make technology interoperable, able to interact and pass information between systems. For example, the Internet is only as powerful as it is because of the huge number of connections it has and the messages it can pass between the countless machines it stitches together. When you ask Siri what is the population of the world is and your iPhone gets the result via the Wolfram|Alpha service, or when you use Google Maps and it shows you how much an Uber might cost, that's interoperability. But when we make systems interoperable, we risk incurring massive amounts of complexity as a downside. We are now not just building interconnected systems—such as each of the individual machines and devices that make up the Internet—but interconnected systems of systems.

In addition to this kind of interoperability, we are also building interdependence between different kinds of technologies, such as the Internet coupled with the power grid. Researchers have studied these sorts of systems to understand their strengths and weaknesses and have found that certain types can and will fail under many conditions; for example, a cascade can start if only a relatively small fraction of the power grid goes down. One response would be to pull back on making these interconnections between technological systems. But that's nearly impossible. The cost of construction of interconnected systems is too low: in an age of interoperability, where engineers and designers purposefully create interfaces for each system, it's *too easy* to build and connect systems together.

When we build something new, there is a tradeoff between the cost of failure and the cost of construction. The cost of failure is how bad it will be if something goes wrong. If your word processor application crashes, you lose any unsaved work. Nobody really wants that outcome, but it's a relatively low cost of failure. However, if a problem in the electrical grid renders a large portion of the United States without power, that's an extremely costly failure. For example, the Northeast Blackout in 2003 affected 50 million people, contributed to the death of eleven people, and cost an estimated $6 billion.

Each failure—and its cost—can be balanced against the cost of building a system. Historically, the more important systems we have constructed have cost a lot to build; it takes more resources to construct the infrastructure for a banking system than for a chat program. Therefore, it has been worthwhile to ensure that costly systems are resistant to failure (which often adds yet more to the cost of construction). In other words, the high cost of construction has made it vital that the cost of failure be reduced through lots of checking and effort.

For a long time, this approach worked: the cost of construction appeared to overweigh the cost of failure, with important systems that we rely on for the vital functions of society costing a lot of money to build. But things have begun to change. The cost of construction has gone down drastically, thanks to off-the-shelf tools and components, resources available in the cloud, and much more. Tech start-ups no longer need much initial funding: you can build and market test a sophisticated tool quickly and cheaply.

Simultaneously, and thanks to some of the same trends, the cost of failure associated with interconnection has gone way up. It has now become easy and cheap to make the types of interconnected systems that incur huge costs when something goes wrong. When digital maps are connected to software that provides directions, small errors can be disastrous (for example, Apple Maps mislabeled a supermarket as a hospital when it was first unveiled). In an age when we can conceive of synthetically generating microbes by sending information over the Internet, the risk of some sort of biological disaster grows much higher. The poliovirus has been reconstructed in a lab using mail-order biological components; there are now start-ups working to allow biology experiments to be run remotely; and it is not hard to imagine, in our increasingly automated world, that a biological agent generated by software could inadvertently be unleashed upon the world. When the cost of construction plunges and the cost of failure rises, we enter a realm of technological complexity that should give everyone pause.

In general, interaction within and between systems increases, which increases the complexity of our overall systems. This increased interconnection is virtually a basic imperative of technology, according to some. Technology ultimately connects, interacts, and converges. And when it does, it acts as a further force moving us toward complication.

While these trends have been happening for as long as we have been building technology and large systems, they've become more powerful in recent years. As noted in this book's introduction, the computer scientist Edsger Dijkstra recognized the radical novelty of our large systems, specifically our computational ones. In 1988, Dijkstra noted that programming a computer requires the ability to jump across massive differences in scale, something no one had really had to handle before computers. To understand this traversal of such a massive hierarchy, Dijkstra gave the example of intellectually navigating from the scale of a single bit in a program or machine up to several hundred megabytes of storage, leaping between the very small and very large. This involves jumping up nearly a billion times in size, a change in scale far more extreme than anything anyone had grappled with before. This has only become more extreme since: everyday users of technology must now be familiar with prefixes of giga- or tera-, prefixes that make us responsible for such huge differences in scale that they border on the astronomical.

Only in the past several decades have large systems become so big and so interconnected that we have found ourselves with, to use Dijkstra's phrasing, "conceptual hierarchies that are much deeper than a single mind ever needed to face before."

However, even if we could prevent our systems from accreting and interacting, there's another reason our systems become more complex over time. And it's one that we are even less able to untangle.

The Edge Case

Imagine you want to create a calendar software application. Sounds straightforward, right? In many ways it is. The calculation of the number of days in the year is relatively easy. It's 365, unless it's a leap year, and then we add a single day to the end of February. But how do we calculate a leap year? Well, you look at the year. If the year is divisible by 4, but not by 100 (unless it's also divisible by 400), then you get a leap year. Easy. Sort of.

But you also want this program to handle changes in time zones, right? Well, that shouldn't be hard. You simply use the geographical coordinates from GPS and determine which time zone you are in. You compile a list of the states according to their time zones. However, time zones don't always divide neatly along state boundaries. Now you need something even more fine-grained, down to fairly small regions. Also, don't forget Arizona, a state that for the most part doesn't use daylight saving time, placing it in its own strange temporal realm.

Do we need our calendar to handle holidays, too? Of course, and since those are straightforwardly defined they should be easy to add. Well, they are, most of the time. Thanksgiving is the fourth Thursday in November and Veterans Day is always November 11. But what about Passover? It seems we now need to make sure the calendar app can integrate with another calendar that is based on lunar information, since Passover starts the evening of the fifteenth day of a lunar month, as defined by the Hebrew calendar. So we need a bit more information to be included than we initially anticipated.

Do you want the calendar to handle other time periods, and be accurate in the past as well as the future? If we go back into the nineteenth century, then individual towns each had their own times, prior to standardized time zones, and this information might need to be hard-coded into our app as well. Similarly, recall that over the course of several centuries there was a flurry of switching from the Julian calendar to the Gregorian calendar across the globe, but the more accurate calendar was adopted unevenly. For example, the reason the Russian October Revolution is commemorated in November is that at the time the revolution occurred, Russia was still using the Julian calendar, and its dates differed by more than a week from parts of the West; the revolution happened at the end of October according to the Julian calendar. If we want to be thorough, we might want all this information in our calendar as well.

We could go on.

Systems we build to reflect the world end up being complicated, because the world itself is complicated. It's relatively straightforward to handle the vast majority of the complexity through a simple model, such as knowing that the calendar is either 365 days or 366 days, with a simple algorithm to keep track of when it's which. But if you crave accuracy—whether in making sure you never miss an appointment, or in building a self-driving vehicle that not only won't get lost but also won't injure anyone—things suddenly become a good deal more complicated.

These kinds of complications are known as *edge cases:* the exceptions that nonetheless have to be dealt with, otherwise our technologies will fail. Edge cases range from the problem of the leap year to how to program database software to handle people's names that have an apostrophe in them. Edge cases are far from common, but they occur often enough that they must be acknowledged and managed properly. But the process of doing so robs our technologies of simplicity and makes them much more complex. This can be seen quite clearly in the ways some of our scientific models—which are a type of technology—have changed over time. We turn to an example from the social sciences: linguistics.

Common Rarities

My middle school English teacher taught me grammar. At a young age, I memorized how to conjugate the irregular verb "to be," as well as a list of prepositions in the English language, and learned how to diagram sentences. A sentence diagrammed—stripped to its logical skeleton—is magnificent to behold. You reduce language to its atomic features, like nouns, verbs, and adjectives, and show how they all hang together.

Though languages do not adhere to a set of equations, grammar maintains a distinct sort of beauty and order. Nonetheless, building a system to process language is not an easy task. Languages have idiomatic expressions, words whose connotations are often far more slippery than we would like, and an informality that makes grammar a rule set that is more nodded to than obeyed. These complications are all edge cases of a sort, the kind that prevents us from building a simple rule on the assumption that every sentence is some variant of Subject–Verb–Direct Object. We can better understand these edge cases of language by looking at things known as *hapax legomena*, what we might call common rarities.

Ever used the word "snowcrie"? I doubt it. In fact, "snowcrie" doesn't even have a definition. As far as we know, it was a typo of sorts. According to the *Oxford English Dictionary,* the word occurred in a line in a poem from 1402: "Not in Goddis gospel, but in Sathanas pistile, wher of sorowe and of snowcrie noon is to seken." Scholars think it might have been an error, likely meant to be "sorcerie."

Whatever its true nature, "snowcrie" is what is known as a *hapax legomenon,* a word that only occurs once in a given corpus—a massive, often complete, collection of texts—such as from an entire language or time period. In this case, the corpus consists of everything in English from printed sources available to the dictionary editors. But the body of text doesn't have to be so large. Within the Shakespearean corpus—all the writings of William Shakespeare—there are numerous hapax words, such as "honorificabilitudinitatibus," which essentially seems to mean "of honor."

When a corpus is all (or nearly all) we have of an entire language, such as the Hebrew Bible in the case of biblical Hebrew, hapax words can sometimes be quite vexing, since we might have little idea of their meaning. But hapax legomena aren't strange statistical flukes or curiosities. Not only are they more common as a category than we might realize, but their existence is related to certain mathematical rules of language. The frequency of words in a language is described by what is known as a *power law* or, more commonly, a *long tail.* These types of distributions, unlike the bell curves we are used to for such quantities as human height, have values that extend far out into the upper reaches of the scale, allowing both for exceedingly common words such as "the" and for much rarer words like "flother."

Often about half of the words in a corpus turn out to have only a single occurrence, making them hapax legomena. They occupy the "long" part of the long tail. While it is rare that you will encounter a specific hapax word, it is likely you will encounter them as a category. To translate this into the world of movies, it's rare to find someone who has seen *The Adventures of Buckaroo Banzai Across the 8th Dimension,* but it's not rare to find someone who has seen at least one weird cult film.

Rare words, then, are quite important as a group; they permeate our language. If we're building a computer program to model language, it's tempting to abstract away rare words or odd grammatical structures as outliers. But as a category, if not individually, they make up a large portion of language. Abstracting them out will cause our model to be woefully incomplete. To avoid losing our exceptions and edge cases, we need models that can handle the complexity of these exceptions and details. As Peter Norvig, Google's director of research, put it, "What constitutes a language is not an eternal ideal form, represented by the settings of a small number of parameters, but rather is the contingent outcome of complex processes."

So, computational linguists incorporate edge cases and try

to build a robust and rich technological model of a complex system—in this case, language. What do they end up with? A complex technological system.

For a clear example of the necessary complexity of a machine model for language, we need only look at how computers are used to translate one language into another. Take this great, though apocryphal, story: During the Cold War, scientists began working on a computational method for translating between English and Russian. When they were ready to test their system, they chose a rather nuanced sentence as their test case: "The spirit is willing, but the flesh is weak." They converted it into Russian, and then ran the resulting Russian translation back again through the machine into English. The result was something like "The whiskey is strong, but the meat is terrible."

Machine translation, as this computational task is more formally known, is not easy. Google Translate's results can be imprecise, though interesting in their own way. But scientists have made great strides.

What techniques are used by experts in machine translation? One early approach was to use the structured grammatical scaffolding of language I mentioned above. Linguists hard-coded the linguistic properties of language into a piece of software in order to translate from one language to another. But it's one thing to deal with relatively straightforward sentences, and another to assume that such grammars can handle the diversity of language in the wild. For instance, imagine you create a rule that handles straightforward infinitives, but then doesn't account for split ones, such as "To boldly go where no one has gone before." And what about regional phrases, like the Pittsburghese utterance "The car needs washed" (skipping over "to be")? The rules will cower in fear before such regionalisms.

Using grammatical models to process language for translation simply doesn't work that well. Language is too complex and quirky for these elegant rules to work when translating a text. There are too many edge cases. Into this gap have stepped numerous statistical approaches from the world of machine learning, in which computers ingest huge amounts of translated texts and then translate new ones based on a set of algorithms, without ever actually trying to understand or parse what the sentences mean. For example, instead of a rule saying that placing the suffix "-s" onto a word makes it plural, the machine might know that "-s" creates a plural word, say, 99.9 percent of the time, whereas 0.1 percent of the time it doesn't, as with words like "sheep" and "deer" that are their own plurals, or irregular plurals such as "men" or "feet" or even "kine." Now do similar calculations for the countless other exceptions in the language.

Out of the chaos comes order—but at a price. The most effective translation program ends up being not a simple model but a massive computer system with a large number of parameters, all fit to handle the countless edge cases and oddities of language. These kinds of models "based on millions of specific features perform better than elaborate models that try to discover general rules," in the words of a team of Google researchers. Exceptions must be cherished, rather than discarded, for exceptions or rare instances contain a large amount of information.

The sophisticated machine learning techniques used in linguistics—employing probability and a large array of parameters rather than principled rules—are increasingly being used in numerous other areas, both in science and outside it, from criminal detection to medicine, as well as in the insurance industry. Even our aesthetic tastes are rather complicated, as Netflix discovered when it awarded a prize for improvements in its recommendation engine to a team whose solution was cobbled together from a variety of different statistical techniques. The contest seemed to demonstrate that no simple algorithm could provide a significant improvement in recommendation accuracy; the winners needed to use a more complex suite of methods in order to capture and predict our personal and quirky tastes in films.

This phenomenon occurs in all types of technology. When building software more generally, the computer scientist Frederick P. Brooks Jr. has noted, "The complexity of software is an essential property, not an accidental one."

Even the complex systems that make up the law are subject to the rule of exceptions and edge cases. While we think of the boundary between what is legal and what is not as a clear dividing line, it is far from being so. Rather, the boundary becomes further and further indented and folded over time, yielding a jagged and complicated border, rather than a clear straight line. In the end, the law turns out to look like a *fractal:* no matter how much you zoom in on such a shape, there is always more unevenness, more detail to observe. Any general rule must end up dealing with exceptions, which in turn split into further exceptions and rules, yielding an increasingly complicated, branching structure. The legal scholar Jack Balkin discusses this in an article evocatively titled "The Crystalline Structure of Legal Thought":

> We might consider whether under an objective standard of negligence, there is an exception for children, or a different standard for insane persons, or for those who are blind, or intoxicated, and so forth. This leads us to further rule choices, each of which leads to additional branches of doctrinal development. Assume, for example, that we follow one of these branches of doctrinal development and create an exception for children (which is now the majority rule). We might consider if there is an exception to that exception when the child engages in an adult activity (this, too, is the case now generally). We might then go on to ask if operating a motorcycle is an adult activity within the meaning of that rule, and if

> so, whether operating a motorscooter is also an adult activity. Put together, we have a descending series of rule choices of increasing factual complexity and specificity. . . .

The law professor David Post and the biologist Michael Eisen teamed up to examine this as well, and while they admit they can't prove that a legal statement can always branch further, and that it's "turtles all the way down," they do note that "we have never met a legal question that could not be decomposed into subquestions." Post and Eisen even show through simulations that certain types of branching structures that mimic legal systems actually can have a fractal structure. Testing this, they find features indicative of fractals when looking at actual legal citations of court case opinions. The fractal complexity of the law might be more than an evocative metaphor.

As the scholars Mark Flood and Oliver Goodenough recognize, "Much of the value of good contracts and good lawyering derives from the seemingly tedious planning for all the ways that a relationship might run off the rails." In other words, legal complexity is often derived from exceptions and their complications.

Whichever technological system we look at—whether it be a legal system, a piece of software, an appliance, a scientific model, or whatever else that we have built—each is driven to become more complex and more kluge-like because of exceptions and edge cases, alongside the twin forces of accretion and interaction.

The Imperfect Nature of These Forces

While I've described these forces of increasing complexity as inexorable, they are not always impossible to resist. However, we can see their true strength by looking at our attempts to work against these forces, and how often we fail to eradicate the kluge.

There are some strategies that appear to bring order and logic to our technologies. If we can build and design our systems differently, or modify and rebuild them, perhaps we have a hope of taming these technologies. For example, we could try to uncouple certain systems—breaking them apart into smaller pieces—so things stay relatively simple and manageable. The physics-trained sociologist Duncan Watts, a principal researcher at Microsoft Research, has argued that in the financial realm, one solution to failures that arise from complexity is to simply remove the coupled complications: if a firm becomes too big and its failure is expected to cause a cascading shock, it must be divided up or shrunk.

Similarly, some scholars have spoken of finding the optimal levels of interoperability for a large system. An optimal level would be one that allows powerful systems to operate well without the downside brought by high unpredictability. Optimal interoperability, then, rather than maximum interoperability, is the goal. Of course, this is not so easily achieved. It is one thing to want the right level of interoperability or interaction, and another to know how to build it. One way to do this involves using certain design principles, such as building understandability and modularity into our creations.

As discussed earlier, when a system has a great degree of interconnectivity, it is often difficult to pull apart the pieces and see what's going on. But sometimes there are parts of a large system that are more tightly connected among themselves than they are with other parts. In other words, there are *modules,* parts of a system that are tightly interconnected and reasonably self-contained. We see many modules in biology, with integrated parts that act in concert, on scales from mitochondria to the human heart. These modules are still intimately connected to the rest of the system, whether through other parts of the body or through chemical signals—I do not recommend trying to remove your heart—but they are relatively distinct and can be understood, at least to some degree, by themselves.

We see modularity in technology, too, such as when a piece of software is made up of many independent functions or pieces; or when you can swap out different applications that do the same thing, but in different ways (think exchanging Microsoft PowerPoint for Keynote on the Mac); or when you examine particular, relatively distinct sections of the United States Code of federal legislation. Modularity embodies the principle of abstraction, allowing a certain amount of managed complexity through compartmentalization.

Unfortunately, understanding individual modules—or building them to begin with—doesn't always yield the kinds of expected behaviors we might hope for. If each module has multiple inputs and multiple outputs, when they are connected the resulting behavior can still be difficult to comprehend or to predict. We often end up getting a combinatorial explosion of interactions: so many different potential interactions that the number of combinations balloons beyond our ability to handle them all. For example, if each module in a system has a total of six distinct inputs and outputs, and we have only ten modules, there are more ways of connecting all these modules together than there are stars in the universe.

In some realms that can be heavily regulated, such as finance or corporate structures, our dreams of increasing modularity or finding the ideal level of interoperability might work. We could, for instance, mandate the breaking up of certain institutions if they reach a given size. But in most other types of technological systems, these ramifying interconnections will often continue apace, no matter our desires. With a relatively small system, building it modularly or piecewise is possible, but as things grow, this kind of clear modularity becomes less likely. As a result of social pressures and the legacy structures of these systems, we continue interconnecting and muddling these systems over time. They grow and complicate despite our desire for simplicity.

We can try to build better-designed systems, and for a while that might even work. For example, there are good computer science and engineering practices, essentially "engineering hygiene," that can drastically reduce the complexity of a system, such as avoiding certain types of variables in one's computer programs. If Toyota had followed these practices, the overall complexity of its system would have been far lower. In addition, professional software development includes methods that can reduce the number of bugs in the programs that are created, bringing rates as low as 0.06 defects per 1,000 lines of code (a very low number); and there are specific practices for managing teams when people work together to construct, operate, and maintain a complex technology, helping to reduce many problems we might encounter in our systems. But in the long term, accretion, interaction, and the edge cases often swamp these attempts at simplification.

As our systems become more complex over time, a gap begins to grow between the structure of these complex systems and what our brains can handle. Whether it's the entirety of the Internet or other large pieces of infrastructure, understanding the whole is no longer even close to possible.

But why must this be so? We next turn to the social and biological limits of human comprehension: the reasons why our brains—and our societies—are particularly bad at dealing with these complex systems, no matter how hard we try.

Chapter 3
LOSING THE BUBBLE

In 1985, a patient entered a clinic to undergo radiation treatment for cancer of the cervix. The patient was prepared for treatment, and the operator of the large radiation machine known as the Therac-25 proceeded with radiation therapy. The machine responded with an error message, as well as noting that "no dose" had been administered. The operator tried again, with the same result. The operator tried three more times, for a total of five attempts, and each time the machine returned an error and responded that no radiation dosage had been delivered. After the treatment, the patient complained of a burning sensation around her hip and was admitted to the hospital.

Several months later, the patient died of her cancer. It was discovered that she had suffered horrible radiation overexposure—

her hip would have needed to be replaced—despite the machine's having indicated that no dose of radiation was delivered.

This was not the only instance of this radiation machine malfunctioning. In the 1980s, the Therac-25 failed for six patients, irradiating them with many times the dose they should have received. Damage from the massive radiation overdoses killed some of these people. These overdoses were considered the worst failures in the history of this type of machine.

Could these errors have been prevented, or at least minimized? If you look at a 1983 safety analysis of these machines by the manufacturer, one of the reasons for the failure becomes clear. The individuals involved in designing and testing these machines looked only at hardware errors and essentially ignored the software, since "software does not degrade due to wear, fatigue, or reproduction process." While this is a true statement, it completely ignores the fact that software is complex and can fail in many different ways. This report implies a lack of awareness on the part of its makers that software could have a deadly complexity and be responsible for a radiation overdose. Software bugs are a fact of life, and yet the safety analysis almost completely ignored the risks they present.

The people responsible for ensuring the safety of the Therac-25 misunderstood technological complexity, with lethal consequences. In hindsight it's almost easy to see where they went wrong: they downplayed the importance of whole portions of the constructed system, and the result was a catastrophic failure. However, it's more and more difficult to diagnose these kinds of problems in new technology. No matter how hard we try to construct well-built, logical technologies, there will always be some part that is beyond our complete understanding. The reason for this is simple: we are human. There is a fundamental mismatch between how we think and how complex systems operate; the ways in which they are built make them hard—or impossible—to think about.

One of the first things you learn when programming is to count differently. This doesn't mean counting in binary or even in hexadecimal (16 different digits, rather than the usual 10)—for most programmers, this is an interesting but unnecessary skill. What I mean here is counting from zero, with the first object in a list always being the "zeroth." Computer programmers count from zero rather than one because that's the way machines count. As the writer Scott Rosenberg notes, the space between machine counting and human counting is an area where we make adjustments in computer code, but it's also where errors and bugs originate. We have to adjust counts by one, incrementing or decrementing over and over, in order to adjust for the differences between how humans intuitively number the world and how machines enumerate their variables. When we fail to adjust, the errors multiply.

The fact that we don't count from zero, but our computers

do, is emblematic of the larger rift between human thought patterns and how large systems are constructed and how they operate. We can't keep track of all the parts in these systems, how they all interact, and how each interaction leads to a new set of consequences. Our human brains are just not equipped to encompass this kind of complexity. In this space between how complex systems are built and how humans think, we find the complications that lead to reduced understanding and to unanticipated consequences and problems.

In the military, soldiers are often confronted with complicated situations that require holding a great deal of information in their heads simultaneously, while maintaining the capacity for rapid response. But sometimes a situation is just too complicated, too stressful, and too messy. A soldier gets overwhelmed and loses the capacity to manage the rush of events. When this happens, a soldier is said to "lose the bubble." As Thomas Homer-Dixon describes the experience, "the comprehensible and predictable suddenly become opaque and bewildering." Awareness of the situation and system drops precipitously, and the soldier is left unable to process the barrage of stimuli and act on it.

This is not a problem only for those in tense situations and in scenarios with lots of complexity, such as directing air traffic or reacting calmly in a battle. It has become a problem for all of us, individually and collectively, in coping with our human-made technological systems. We have lost the bubble.

In the Entanglement, we lose the bubble in two related ways: we are unable to fathom the structure and dynamics of huge and complex systems themselves—the way the different pieces interact as a whole; and we are unable to make much headway into the vast quantity of knowledge and the specialized expertise it would take to fully understand how these systems operate. To see more clearly why we think about the world in ways that are ill-suited to complex technological systems, we can again look at language.

When Our Brains Fall Short

Recursion is a computer science term that means, essentially, self-reference: it describes a section of computer code that refers back to itself. It's also spawned some programming humor. Search for the term *recursion* on a search engine and it might ask you, "Did you mean *recursion*?" This is funny to a distinct subset of humanity.

Recursion is also built into the fabric of how we speak. Language has a recursive capability; in fact, it is infinitely recursive, at least in theory. You can say, "He said that the dog is brown," as well as "She thought he said that the dog is brown," the second sentence embedding the first within it. And if you're more daring, you could even utter, "I remembered that she thought he said that the dog is brown." Depending on the structure of the sentence, this embedding of sentences within sentences can be done at the beginning of a sentence, in the middle, or at the end, over and over.

Its recursive nature makes language infinitely rich. Imagine a relatively small language with 1,000 verbs, 10,000 nouns, and a rule that the only sentences one can make are of the form *noun verb noun.* The linguistic capacity of this language is huge: you can make up to 10,000 × 1,000 × 10,000 sentences, which is 100 billion sentences. If you spoke a sentence every ten seconds, it would take more than thirty thousand years to exhaust all possible sentences. What's more, the example above is a particularly impoverished language. To speak every possible sentence in a language even slightly more complicated—with a greater number of words or more complex sentence structures—would take amounts of time that might better be described as geological eons.

While these numbers are inconceivably large, they are still finite. To proceed from mind-boggling finitude to true infinity, we must use recursion. And once you introduce recursion—allowing an arbitrarily large number of clauses to be embedded within one another—a language becomes theoretically infinite in its richness.

But in practice this isn't quite true. It's silly to say that language allows for an arbitrarily large number of embedded clauses: that may be technically feasible according to the rules of grammar, but our brains simply can't parse that much recursion. As much as we would like our languages to be infinite and variegated, we can't handle sentences with a recursion depth of much more than two or three.

Here are some sentences from the linguist Steven Pinker that not only are hard to understand, they don't even look syntactically correct:

The dog the stick the fire burned beat bit the cat.

The rapidity that the motion that the wing that the hummingbird has has has is remarkable.

Each of these has only a small amount of nesting. For example, the first sentence means that the dog—the one that was beaten by a burnt stick—bit the cat. It is constructed by modifying "the dog"—of "The dog bit the cat"—with a description of the stick. This sentence has only two levels of nesting. If you were to go up to ten levels, the sentence would be effectively impossible to make sense of. And if we can't handle ten, we certainly can't deal with numbers that scrape the ceiling of infinity.

Humans can do some types of linguistic processing, such as translation, better than computers can. But for parsing sentences, computers have numerous advantages. While human cognitive processing is limited by our working memory, computers can use large memory stores to put each portion of the sentence in its place, and can then construct the tree of the sentence, rendering it meaningful. For this reason, computers can easily parse sentences that flummox the human mind. For example, the sentence "This is the cheese that the rat that the cat that the dog chased bit ate," although strange and impenetrable to human ears (and eyes), can be parsed by a machine.

There are even computer programs that have these syntactic structures built in, allowing for the creation of quite large sentences. Instead of processing these complicated sentences, they generate random text that appears realistic in the style of certain authors. Consider Kant Generator, which can make such sentences as "Since knowledge of the phenomena is a priori, the reader should be careful to observe that, so far as regards necessity and the things in themselves, the discipline of human reason, so far as I know, can be treated like our judgments." We are far more complicated creatures than this tiny computer program, but we have a lot of trouble breaking such things down and determining whether or not they're nonsensical.

In a similar vein, there are structures known as *garden path sentences,* such as "The complex houses married and single soldiers and their families." These are intriguing sentences that begin one way but end up having a different grammatical structure and meaning than we were initially led to expect. We fill in the expected meaning and are surprised and often momentarily confused when the sentence takes a different route.

Of course, recursion and other grammatical tricks are far from the only such demonstrations of human cognitive limits. There's a whole cottage industry of tasks that we're not particularly good at that machines handle with ease. I'm not talking about perceptual challenges like optical illusions, but bona fide examples of our mental-processing limits. For example, have you ever wondered how long a string of numbers you can memorize at once? For most of us, it's about seven. It's hard to memorize much more than a telephone number, minus the area code.

Or take counting objects, or even dots on a screen. We can definitely count lots and lots of objects and dots, but how many can you count in one glance, perceiving that number immediately? It turns out that this number is very small. When you're looking at a group of dots, your brain ends up grouping them into multiple smaller groups, often of about three or four. Visually, most of us can immediately perceive only four or so items. This ability to perceive at once is called *subitizing,* and it's a weird quirk of our brains that we can do this effectively only for a small number. Just as we have trouble reading long and winding sentences, we have trouble counting more than four objects at once.

In other comparisons with machines, we are also pretty pathetic. It takes about eight seconds to transfer a piece of information into our long-term memory. In less than that amount of time, you can download *War and Peace* to your laptop. And unlike some simple computers, we're pitiful at multitasking. Our neurons are more than a million times slower than a computer circuit, and, according to one estimate, our long-term memory can't hold much more than one of my family's old Macintoshes from the 1980s could.

Are our brains capable of reaching beyond these rather meager limits? Research in human cognition is not encouraging on the matter. Much as computers can be tweaked to work faster than intended—this is known as *overclocking*—we can sometimes soup up our mental engines as well, using pharmaceuticals. But when we study these attempts to "overclock" our brains, we discover trade-offs. Just as overclocked computers can overheat, our brains can also suffer from being pushed beyond their limits. It seems that our brains have been delicately optimized by evolution, and attempts to tinker with them can create serious problems.

You can see examples of such trade-offs if you look at those rare individuals who have unlimited memories—they remember essentially every fact they encounter and every occurrence they witness. But they are not superhumans. In fact, they end up being hampered by such issues as trouble recognizing faces. Because their memory is so detail-oriented, anytime there is a change in how a person appears, it is difficult to recognize that person as the same one. In "Funes the Memorious," a short story by Jorge Luis Borges, the title character is burdened by a perfect and complete memory. Every change and detail generates a new memory. Just as it was for the fictional Funes, in real life an unbelievably good memory seems to cause problems with skills such as abstraction, leaving one burdened with huge amounts of unnecessary information.

And just as there are outliers in cognitive processing—such as these individuals with prodigious memories and those who can calculate huge arithmetical operations in their heads—we also see extremes of insight in understanding something complex. For example, take the mathematician Srinivasa Ramanujan. A self-taught genius who worked during the early part of the twentieth century, Ramanujan was not your average mathematician who tried to solve problems through trial and error and occasional flashes of brilliance. Instead, equations seemed to leap fully formed from his brain, often mind-bogglingly complex and stunningly correct (though some were also wrong).

The Ramanujan of technology might be Steve Wozniak. Wozniak programmed the first Apple computer and was responsible for every aspect of the Apple II. As the programmer and novelist Vikram Chandra notes, "Every piece and bit and byte of that computer was done by Woz, and not one bug has ever been found. . . . Woz did both hardware and software. Woz created a programming language in machine code. Woz is *hardcore*." Wozniak was on a level of technological understanding that few can reach.

We can even see the extremes of our brain's capacity—as well as how its limits can be stretched—in the way London cabdrivers acquire and use what is known as The Knowledge. The Knowledge—a wonderfully eldritch term—is the full complement of all details of the metropolitan London area: the 25,000 roads and their interconnections, as well as parks, landmarks, statues, restaurants, hotels, and every other conceivable detail that a cabdriver must know in order to accurately and efficiently transport a passenger from any one location to another. Learning The Knowledge, and being certified as a cabdriver, can take several years of intense memorization and exploration of London. The result, though, is that the brains of these cabdrivers visibly change: the posterior hippocampus, a region important for spatial memory, increases in size.

But even these outliers, impressive as they are, have limits. Ramanujan still got things wrong, and Wozniak would certainly also run up against cognitive limits, recursive or otherwise. And London cabdrivers would be hard-pressed to contain the entire Earth's road network in their minds.

Besides our limited memory storage and retrieval capacities and our ability to hold only so much in our conscious minds at once, we have difficulty grasping the implications of interconnections within a system. Specifically, we are confounded by nonlinear systems. When something changes in a linear way—a small change creating a small difference, a bigger change yielding a bigger difference—we are essentially tasked with extrapolating a straight line. Our brains have little difficulty doing this, because a linear system's inputs are directly proportional to its outputs. But when a small cause caroms through a large interconnected system and results in a big effect—so that the system is changed in a disproportionate way—we are unprepared for this highly nonlinear result.

A nonlinear system's behavior is modulated by feedback and the magnification of inputs (or even the opposite: a big value giving you a tiny effect), making it much more difficult to relate the inputs to the outputs. We are no longer extrapolating a straight line; the variables interact in swooping and complicated curves, over which our brains stumble. These shortcomings cause us to have difficulty grasping complex systems, even those we have built ourselves.

Too Complex to Handle

Philosophy is a large field, with specialties ranging from political philosophy and ethics to the philosophy of science and technology. Within the philosophy of technology, there is a growing interest in the philosophical implications of software: How should we think about our computational creations? The philosophers John Symons and Jack Horner at the University of Kansas have examined how our construction of software—one type of technological system—can yield incomprehensibility almost immediately.

The simplest reason for this is branch points. If you have a piece of technology that does one thing if condition A is true but something else if condition B is true, this is considered a branch point, or an if-then statement, in the parlance of programmers. For example, a computer program might add ten to a number if that number is odd, but only five if that number is even.

As Symons and Horner note, once a computer program incorporates these branch points, the number of potential paths the program can take when run on a machine begins to multiply. Using some reasonably conservative calculations, they show that a program of only 1,000 lines (relatively short for even pretty simple programs, and much shorter than most programs used in "the wild") already has 10^{30}—more than a trillion trillion—potential pathways that can be traversed, assuming that branch points occur every so often in the computer code. To check all possible paths—understanding the implications and soundness of each one—is not only infeasible, it is impossible. This system is not simply difficult to understand; it is effectively hopeless to fully understand, in all its details, within the age of the universe. In other words, the vast majority of computer programs will never be thoroughly comprehended by any human being. This includes programs on your laptop, computer code in your kitchen appliances, and software that determines how airplanes are directed around the globe.

Of course, computer programs can still be understood, at least on some level, without manually traversing every potential path. That is one of the features of abstraction, as discussed in the first chapter. Abstraction, combined with various rigorous methods of testing, reduction of errors, and software "hygiene"—such as not using GOTO or certain types of variables—can reduce our lack of understanding. But we will never be truly sure that we know all implications and potential situations. Users, whether scientists working with a model of the world, technicians operating a large machine, or drivers of state-of-the-art automobiles, must be satisfied with incomplete understanding as part of living in the Entanglement.

We also see the effect of large numbers of components and interconnections when we encounter analyses based on large datasets, where huge quantities of data points are fed into algorithms that provide us with predictive power, but sometimes at the expense of human meaning. Google recently turned powerful computional methods on itself, seeking to boost the energy efficiency of its data centers by feeding a slew of their properties into a computer model, including everything from the total number of condenser water pumps running to outdoor wind speed. To quote Google's blog, "In a dynamic environment like a data center, it can be difficult for humans to see how all of the variables . . . interact with each other. One thing computers are good at is seeing the underlying story in the data, so [a data center engineer] took the information we gather in the course of our daily operations and ran it through a model to help make sense of complex interactions that his team—*being mere mortals*—may not otherwise have noticed." (Emphasis mine.)

It's very difficult to follow the mathematical details of these kinds of massive technological models. But as Douglas Heaven, the chief technology editor at *New Scientist* magazine, has written, even if we are able to, it wouldn't necessarily be meaningful to us. A choice or an answer produced by such a piece of software is not arrived at the way we would, and often cannot be understood in terms of a statable general rule or idea. Rather than a straightforward path of logic, the decision is based on an enormously complex set of calculations. We throw in huge amounts of information and data and let the massive piece of software churn something out. We get an answer, and it works, but we are missing insight into the process by which it came to be an answer. In area after area, from the law to the hardware we build, we are partnering with computers to help navigate incredibly complicated technologies. But in the process, we find ourselves largely mystified by how these systems we depend on operate.

This process is only accelerating. One computational realm, evolutionary computation, allows software to "evolve" solutions to problems, while remaining agnostic as to what shape the eventual solution will take. Need an equation to fit some data? Take a page from biology. Create a population of potential solutions within a computer program and allow them to evolve, recombining, mutating, and reproducing, until the fittest solutions emerge triumphant. An evolutionary algorithm does this splendidly—even if you can't understand the final answer it comes up with.

A number of years ago, research was conducted to design a type of computer circuit. A simple task was created that the circuit needed to solve, and then the researcher tried to evolve a solution in hardware, with candidate circuits mingling in a Darwinian stew. After many generations, the program found a successful circuit design. But this design had a curious feature: parts of it were disconnected from the main circuit yet were somehow still vital to its function. The evolved circuit had taken advantage of weird physical and electromagnetic phenomena, which no engineer would ever have thought of using, to make the circuit complete its task.

In another instance, an equation was evolved to solve another problem, and the result was also recognized as impenetrable. Kevin Kelly, in *Out of Control,* describes it thus: "Not only is it ugly, it's incomprehensible. Even for a mathematician or computer programmer, this evolved formula is a tar baby in the briar patch." The evolved code was eventually understood, but its way of solving the problem appeared to be "decidedly inhuman." This evolutionary technique yields novel technological systems, but ones that we have difficulty understanding, because we would never have come up with such a thing on our own; these systems are fundamentally different from what we are good at thinking about.

In the realm of logistics, powerful algorithms have been developed that route delivery trucks in seemingly illogical ways, leaving drivers dissatisfied with the counterintuitive routes they are being given. And in chess, a realm where computers are more powerful than humans and are able to win via pathways that the human mind can't always see, the machines' characteristic game choices are known as "computer moves"—the moves that a human would rarely make, the ones that are ugly but still get results. As the economist Tyler Cowen noted in his book *Average Is Over,* these types of moves often seem wrong, but they are very effective. When IBM's Deep Blue was playing Garry Kasparov, it made a move so strange that it "was both praised and panned by different commentators," according to one of Deep Blue's builders. In fact, this highly odd but potentially brilliant move was eventually found to be due to a bug. Computers have exposed the fact that chess, at least when played at the highest levels, is too complicated for us, with too many interacting parts for a human—even a grandmaster—to keep in view. We sometimes can't even tell when a decision is flawed.

What about the law, from contracts to regulations? There, too, we see this problem intensifying. Mark Flood and Oliver Goodenough, quoted in the previous chapter, recognize that "[w]hen interpretation is necessary, legalese, even when not ambiguous, makes slow, tortuous reading, with the need to check and recheck the definitions, cross-references, exceptions, etc. in which the complexity is embedded. Lawyer brains, the computational mechanism of traditional contract interpretation, are expensive and subject to cognitive limitations." We should not be surprised to learn that when forty-five tax professionals were given data on a hypothetical family's income, they came up with forty-five distinct conclusions about how much that family should pay in taxes.

Even as the sheer number of components and connections in complex systems overwhelms our processing capacity and causes us to lose the bubble, there is another factor in our reduced comprehension: the limits to how much knowledge—not just memorized raw data, but specialized technical expertise—we can keep in our heads. As technologies draw on more and more different domains of knowledge, even experts lose the ability to know them all.

To address the limits of our knowledge, we have to look at humanity's pursuit of specialization. As we do, it should become clear that this new era of incomprehensibility is not entirely novel. Rather it is a continuation—though an extreme endpoint—of processes that have been occurring for much of human history.

The End of the Renaissance Man

I have on my shelf three books that have the phrase "The Last Man Who Knew Everything" as their title or subtitle. One is an edited volume about Athanasius Kircher, a German Jesuit priest who lived during the seventeenth century. He is viewed today as both a brilliant eccentric and probably a bit of a charlatan, who wrote about everything from astronomy to Egyptian hieroglyphics to a musical organ made from live cats. The second book is about Thomas Young, who was born in 1773 and studied such topics as physics, medicine, and linguistics. And the third concerns Joseph Leidy, born in 1823, a Philadelphia-area paleontologist and naturalist.

Which one was the Last Man Who Knew Everything? I don't know. In fact, there probably wasn't any one human being who ever knew everything we have generated as a civilization. But over the past few centuries, there has been such an explosion of knowledge that it would be remarkable, and almost impossible, to have even a passing awareness of all the new knowledge that was being generated. Our understanding of the universe has become complicated. But before the time of these

Last Men passed, some people did try to understand everything around them. This understanding often involved the embrace of something known as a cabinet of curiosities.

Cabinets of curiosities, or *wunderkammers,* were crammed rooms proclaimed to contain the whole world of knowledge. Collections of sundry and bizarre objects, from stuffed and mounted animals to herbs and paintings, they were generally owned by wealthy and noble Europeans. Cabinets of curiosities were markers of social status, but also windows to understanding our universe and the wonders it was revealing to us. Collectors reveled in the man-made—musical instruments and weapons of war—as well as the natural—skeletons and minerals; their cabinets proclaimed the diversity of the world, unified only by the fact that all could be contained within these collections. As the writer Philip Ball notes in his book *Curiosity,* "The ideal collection was *comprehensive*—not in that it contained an example of every object or substance in the world (although efforts were sometimes made towards such exhaustiveness), but in that it created its own complete microcosm: a representation of the world in miniature."

With a cabinet of curiosities, you could take in the entirety of the universe and all its complications at a glance. But not only did some of these cabinets appear to be little more than a miscellaneous hodgepodge; it was soon realized that they could never be big enough. Ball quotes a point made by the writer Patrick Mauries: that after the discovery of the Americas there was too much diversity to be contained within a single collection. The world was beginning to be recognized as too various and complex. Now, choices had to be made: What should make it into these rooms, and what could be ignored?

Cabinets of curiosities persisted for a long time, in their own way. A couple of decades ago I visited the Niagara Falls Museum. Now closed, it was owned at the time by a friend's father and was one of the last of the wunderkammers. The Hall of Freaks of Nature had example after example of stuffed mutant creatures, from a five-legged cow to a two-headed sheep to another sheep that was nothing more than just two joined heads. Gazing at these mutants, and cases of mounted insects, and Egyptian mummies, I was given a sense of the breadth of the world—and its sheer weirdness.

But not everything in the world was on display in these wunderkammers. That would have been impossible. Choices had been made. In these choices, we can detect hints of a growth in specialization. As knowledge grew beyond the bounds of any one continent, or culture, or mind, to have a confident grasp of the systems around us we would have to specialize—to understand a small field very well, say, advanced weaponry, or some subfield of science. But of course this didn't happen all at once.

For a period of time several centuries ago, there were numerous individuals who attempted to truly make sense of what was around them rather than simply collect everything—and who were still well-versed in discipline after discipline. One example was the philosopher, scientist, and mathematician Gottfried Leibniz, who lived during the seventeenth and eighteenth centuries. According to the scholar Daniel Boorstin, "Before he was twenty-six, Leibniz had devised a program of legal reform for the Holy Roman Empire, had designed a calculating machine, and had developed a plan to divert Louis XIV from his attacks on the Rhineland by inducing him to build a Suez Canal." In the words of Frederick the Great, Leibniz was "a whole academy in himself." Similarly, Isaac Newton stitched together a whole host of phenomena—from how objects fall to the orbit of Mars—through his theory of gravitation.

Around the same time, Gresham College, one of the oldest colleges in England, and devoted to providing public lectures on various topics, had a small faculty in areas such as astronomy, geometry, and music. But in reality, Gresham titles were relatively meaningless; some professors chose their titles based on the quality of the rooms they could get rather than their area of expertise. Specialization was far from anyone's mind during this time. As the mathematician Isaac Barrow noted, "He can hardly be a good scholar, who is not a general one."

But knowledge has grown far beyond any single person's capacity to master it. To build models of the world and new technological systems at the frontiers of what we know, we have had to learn "more and more about less and less"—to specialize in specific domains. Benjamin Jones of Northwestern University has developed a theory about the "burden of knowledge": the idea that to make advances at the frontier of knowledge, you must know a substantial amount of what has come before you. As our collective knowledge has grown over time, this burden has grown, with ever more required to be learned in order to make novel contributions. In one article coauthored by Jones, the authors note that John Harvard, whose bequest, including his private library, gave him naming rights to Harvard University in 1639, donated only 320 books to the university. There are now more than 36 million books and other print materials in the United States Library of Congress. The burden of knowledge weighs ever heavier upon us. As we attempt to build and understand complicated systems, we are required to know more, but also to have increasingly specialized expertise.

The biologist E. O. Wilson described the change thus:

> In 1797, when Jefferson took the president's chair at the American Philosophical Society, all American scientists of professional caliber and their colleagues in the humanities could be seated comfortably in the lecture room of Philosophical Hall. Most could discourse reasonably well on the entire world of learning, which was still small enough to be seen whole. Their successors

> today, including 450,000 holders of the doctorate in science and engineering alone, would overcrowd Philadelphia. Professional scholars in general have little choice but to dice up research expertise and research agendas among themselves. To be a successful scholar means spending a career on membrane biophysics, the Romantic poets, early American history, or some other such constricted area of formal study.

Not only is knowledge itself expanding and bifurcating, but the numbers and the specialization of scholars have also greatly increased.

This puts us in a difficult position. Specialization is required in order to understand more and more about the intricate systems around us, such as the human body, now divided up among numerous specialties in medicine. But at the same time, the systems we are building—the technologies that run our world—are not only intricate and complicated, but also stitch together field after field. We have systems in the world of finance that require an understanding of physics; there are economists involved in the development of computer systems. The design of driverless cars is a good example, requiring collaboration among those with expertise in software, lasers, automotive engineering, digital mapping, and more.

In other words, even as specialization aids us in making advances, we are ever more dependent on systems that draw from many different areas, and require an understanding of each of these. Yet a single person can no longer possess all the necessary knowledge. To any one person, these systems as wholes are truly incomprehensible.

One solution is the growth in multidisciplinary teamwork: build a team of individuals with deep expertise in different areas, and you can make advances at the boundaries and build astonishingly powerful complex systems. In software, while some pieces of technology are built by a single person or a small team, more often it takes large numbers of people, who enter or leave a project or team, contributing to its development over long stretches of time. If you attempt to visualize these patterns of teamwork—and impressive infographics have been created, trying to show how key pieces of software were developed—you will find yourself staring at what look like convoluted bundles of strings, meeting and branching as individuals join a software collaboration, work on different files together, and leave again. It is unsurprising, then, that the products of these processes are not only very complex, but often so complicated that few fully understand them: the person who might know all about a specific feature may be long gone.

Specialization is a successful process that yields impressive technologies, but it, too, leads us into the Entanglement, where we are dependent on knowledge of complex technological systems that we as individuals do not have, and that, in fact, no one may have. There are ways of trying to overcome this predicament: for example, perhaps it's time to bring back generalists and polymaths, inviting them to flourish anew in this modern era—a possibility we will reexamine later in this book. But for now, it is enough to recognize the fundamental clash between the amount of knowledge any individual has the capacity to process and the amount we need to know about the interlocking systems our lives depend on.

Unfortunately, we often fail to recognize this mismatch until it is too late. We build massively complex technologies, secure in the belief that they are constructed on a logical foundation, until they confront us with unexpected behavior: the bugs and glitches that send major systems such as global finance into a tailspin. These unexpected behaviors—the kind that even the creators of these systems have trouble anticipating—can be viewed as technological werewolves. In the imagination of the computer scientist Frederick Brooks, software projects have a tendency to morph into unmanageable monsters: "Of all the monsters who fill the nightmares of our folklore, none terrify more than werewolves, because they transform unexpectedly from the familiar into horrors."

The werewolves of our time are the unexpected behaviors that lurch forth from the systems we build, the sinister embodiments of all the forces that have made our systems ever more complicated and less understandable. We turn to these next.

Chapter 4
OUR BUG-RIDDEN WORLD

Back in the 1980s, the video game Galaga was popular. A classic shooter in which your trusty spaceship had to eliminate all the bad guys, it was one of those archaic video games with simple graphics and goofy sounds—but it also had an intriguing glitch. Early on in the gameplay, if you eliminated nearly all enemies and then avoided those that remained for around fifteen minutes, the baddies would never shoot at you again. A curious situation, but one that could be nicely exploited for some satisfying high scores.

Why did this happen? It seems that the part of the code that held the "shots" misbehaved under certain conditions and neglected to refresh. Some speculate that this was an intentional feature of the game, to allow its developer to enter an arcade and rack up high scores. While that would make a great story,

I'm not sure how likely it is, given the cleaner hidden features and cheats found in other games. It was probably just a bug.

This bug and others like it are signs of the fact that we do not fully understand the systems we build. It requires significant effort to understand what was going on inside Galaga, particularly when something went wrong. Even though the graphics and gameplay were simple, the true level of complexity might only become clear to players when it failed. Bugs are not just annoyances to be fixed. Bugs are how we realize that we are in the Entanglement.

In 1950, Alan Turing noted that machines can and will yield surprises for us in their behavior. And it seems that these surprises will only increase in frequency. As we build systems that become more and more complicated, there is a greater divergence between how we intend our systems to act and how they actually do act. This unexpected behavior is a symptom of exactly what we have been looking at so far: increased complexity and decreased understanding, based on the shortcomings of our brains discussed in the last chapter. This means all of us, not just the tech-challenged end user: even experts are caught by surprise when, say, a rocket self-destructs soon after launch, or carefully crafted pieces of legislation turn out to clash. Bugs and glitches are the unexpected and unwanted by-products of the complexity of our technology. They are not only inevitable whenever a system is complex enough; they are the first hints that we are inching ever closer to complete dependence on systems that we don't understand well at all. These technological werewolves are the heralds of the Entanglement.

The philosophers John Symons and Jack Horner, mentioned in the last chapter, have studied software development, focusing on aspects of its construction that are often left unexamined, or at least examined far less than they should be. Unfortunately, these are the very aspects—such as how numbers get stored and rounded when performing calculations—that can lead to a profusion of problems but are often glossed over or ignored by software builders. For example, in a widely used simulator of gravitation, the kind of computer program used in astronomy research, many errors were found involving one way that numbers are handled. Specifically, in this program there are about 10,000 instances of this kind of mistake in a piece of software only 30,000 lines long. Fixing these errors can actually yield different simulation results.

Over and over, we see these problems arise in our technologies as a consequence of increasing complexity. When a piece of technology becomes complicated, the werewolves appear when we least expect them, with unpredictable effects. These bugs range from fun and strange ones, like the glitch in Galaga, to ones as potentially devastating as the Heartbleed vulnerability, a mistake written into encryption software that may have compromised the security of as many as two-thirds of the websites online, including Facebook and Google, for more than two years.

Then there's Microsoft Windows. In 1996, a computer "bug detective" published *The Windows 95 Bug Collection*—an entire book devoted just to detailing bugs related to Windows 95 and potential workarounds. Some solutions this book suggests for various bugs are easier to manage than others. For instance, one edition of a particular Windows program displays an error message when starting up. The solution? Just ignore the message. But others require a more drastic intervention. For example, if you have a specific controller card, it may not work properly with some computers running Windows 95. What does Microsoft suggest? "[U]se a slower computer or a different Bernoulli controller card." In other words, this problem is not being dealt with and may not even be understood at all.

This is what happens when a system becomes large enough. It interacts with the user, other systems, and itself in unexpected ways. In fact, it is estimated that when a software project becomes twice as large, the *rate* of errors increases. You don't just get twice as many problems, you get more than twice as many problems: the number of errors per thousand lines increases. So a program of 10,000 lines, say, could have four times as many errors as one with 5,000 lines.

This unpredictability and fragility is actually a hallmark of the complex systems that we build. While complicated systems are often incredibly robust to shocks that are anticipated—that is, ones they have been designed for—their complexity can be a liability in the face of the unanticipated.

A mathematical model has even been devised for understanding this specific situation, known as *highly optimized tolerance.* Our systems are optimized to tolerate a wide variety of situations, but anything new can lead them into a catastrophic spiral of failure. Take the Boeing 777, a massive machine of an airplane that contains 150,000 subsystem modules designed to ensure it flies well and adapts to numerous situations. But it can't handle every possible contingency. According to two scientists, "The 777 is robust to large-scale atmospheric disturbances, variations in cargo loads and fuels, turbulent boundary layers, and inhomogeneities and aging of materials, but could be catastrophically disabled by microscopic alterations in a handful of very large-scale integrated chips or by software failures." In other words, as a system becomes more complex, the tiniest stimulus could potentially be a catastrophic disruption. We simply don't understand what might happen.

In fact, these unexpected consequences are related to the edge cases and exceptions discussed earlier. The world is complicated, necessitating a complicated system to handle it. But many of these complications are only rarely encountered. As rare as they are individually, though, these uncommon situations can cause problems for technological systems because there are too many of them to test for properly. For example, returning to vehicles with their sophisticated software, it isn't possible to actually test them exhaustively. As the computer scientist Philip Koopman has noted, "Vehicle testing simply can't find all uncommon failures." Even a human lifetime is not long enough to examine everything.

When the world we have created is too complicated for our humble human brains, the nightmare scenario is not Skynet—the self-aware network declaring war on humanity—but messy systems so convoluted that nearly any glitch you can think of (and many you can't) can and will happen. Complexity brings the unexpected, but we realize it only when something goes wrong.

The Symptoms of the Entanglement

Kate Ascher, a professor of urban development at Columbia University, has released a series of books about how cities, transportation networks, and individual buildings work, and the complexities underlying their respective systems. The books are lush with diagrams and fascinating in their details. They are also somewhat overwhelming. These systems have accreted over decades, sometimes centuries, with layer upon layer being added over time, from road networks to the methods by which energy and other necessities are distributed to the buildings of a city. For example, to provide each of our homes and places of business with water is an enormously complex affair. To give a sense of the enormous scale of just the removal of wastewater, this process involves more than 6,000 miles of underground pipes in New York City alone, part of an elaborate construction that handles over one billion gallons of sewage per day.

However, we usually recognize this complexity only when things go wrong. In spring 2010, the population of metropolitan Boston was treated to a crash course in how water is managed and distributed. On the first day of May that year, a water main broke in Weston, Massachusetts, one that carried water from the Quabbin Reservoir. For several days, the residents of many communities (including Brookline, where I was living at the time) were advised to boil their water, as they were now receiving water from backup sources. However, the residents of Cambridge—just across the river and surrounded by the towns affected—were fine, as their water arrived from its own distinct source. While those working for the city were certainly aware of the complexity of the water system and its idiosyncrasies, much of the metropolitan population likely became aware of these facts only once they experienced the system's failure.

Andrew Blum, in *Tubes,* a book exploring the physical infrastructure of the Internet, begins with a similar realization: the sheer interconnected physicality of the Internet—a tangible network that crisscrosses the globe—is revealed to him only when his Internet connection stops working because wiring in his backyard has been chewed on by squirrels.

A truism in the open source software development world is known as Linus's law (after Linus Torvalds, the creator of Linux): "Given enough eyeballs, all bugs are shallow." In other words, with a large enough group of people examining a piece of technology, any glitch—no matter how complicated it is and how difficult it may seem to remedy—can be fixed, because the chances are high that someone will find a way to overcome the problem.

But as our systems become more and more complicated, this may not be true anymore. All the bugs cannot be eradicated: the likelihood that someone will spot—and eliminate—every bug is far lower when we are reckoning with complex and interconnected systems. Furthermore, each fix can, in turn, lead to new problems. This sounds pretty depressing, and in a way it is. But there is at least a partial way out of this funk.

These technological werewolves are not simply markers of the new era we find ourselves in; they can also point us toward a new way of managing our systems. Just as the water crisis around Boston taught many of us where our water actually comes from, examining bugs is one of the few options we have for learning about our world and thriving in the Entanglement.

What Glitches Can Teach Us

Several years ago, Gmail—Google's email service—suffered an outage, and went down for many users for a total of about eighteen minutes. It was eventually determined that a slight error in an update of one piece of Google's software (the software that ensures that processing is balanced so that no part of the system becomes overwhelmed) caused a large number of servers to be considered unavailable, even though they were working properly. This error didn't affect many of Google's other services, but Gmail turned out to require specific data center information, and so it just went down. This cascade was triggered by such a small problem that few might have thought it could cause such a major meltdown. The bug demonstrated a hidden interconnectivity between certain systems that would have remained hidden if the glitch hadn't occurred.

When you debug a piece of technology—that is, attempt to root out errors—you learn the difference between how you expected the system to operate and how it actually works. Sometimes these bugs, whether in automobile software, Internet security, or our urban infrastructure, are simple, easily understood and fixed. At other times they are incredibly frustrating, and can be nearly impossible to diagnose and remedy. But the cataloging of these failures is the first step to learning about a part of the complex system that we're looking at. This *natural history* of our technological world is vital. Just as naturalists go out into the natural world and study it, cataloging its variety and its complexity, we need a similar approach to our technologies.

We are going to continue to need this sort of "technological natural history" more and more. Take another example from programming. Let's say you're thinking of a number between 1 and 100 and I have to guess it. I first ask you if it's greater than 50. If it is, then I ask you if it's bigger or smaller than 75, and I keep on dividing the remaining numbers in half until I find your number. This method is known as *binary search,* because of the division into two groups, and is a highly efficient method for finding what you are looking for in a large sorted list.

Implementations of binary search are found throughout the world of software. Therefore, it was more than a little disturbing to read the title of a blog post from Google in 2006 discussing a bug in many implementations: "Extra, Extra—Read All About It: Nearly All Binary Searches . . . Are Broken."

While code implementing binary search in our software does generally work well, it turns out that many versions can fail under conditions that involve huge amounts of data: "This bug can manifest itself for arrays whose length (in elements) is 2^{30} or greater (roughly a billion elements). This was inconceivable back in the '80s, when *Programming Pearls* [a classic book on computer programming, whose binary search implementation contains this error] was written, but it is common these days at Google and other places," according to the blog post.

Only in today's world of huge datasets could we have discovered this error and learned more about this particular system that we have built. The bug is a window into how this technology actually works, rather than how we intended it to operate.

There are many other instances when unexpected behaviors provide a mechanism for learning how a technology really works. For example, at the beginning of 1982, the Vancouver Stock Exchange unveiled its own stock index, similar to the S&P 500 or the Dow Jones Industrial Average. It was initially pegged to a value of 1,000 points and then, over nearly two years, steadily dropped. Near the end of 1983, it sat at about half its original value. But this didn't make any sense. There was a bull market in the early 1980s; how could the index be declining? Thus began an investigation into what was happening, culminating in the discovery that the calculations of the index were wrong. Instead of taking the index's value and rounding it to three digits, the algorithm responsible for the index's calculation was simply lopping off what came after the three digits. For example, if the index was calculated as 382.4527, it would be stored as 382.452, even though the final 2 in the truncated number should have been rounded up, for a value of 382.453. When this is done thousands of times a day, actual value is lost—in this case, a lot of value. The error was finally corrected in November 1983, when the index closed around 500 on a Friday, only to open the following Monday at over 1,000, with the lost value added back. This deeper problem in calculation—the flaw in the algorithm—was noticed because of an anomaly (in hindsight, not a particularly subtle one): the index was going down while the market was going up in value.

We see similar examples of failures becoming our teachers when companies and governments use sophisticated machine-learning algorithms to comb through huge datasets. For the most part, these learning systems remain impenetrable black boxes for both the general public and, increasingly, the people working with them. Occasionally, however, these huge systems spit out strange or even worrisome results that inadvertently provide us with a hint of what is happening inside the system.

Take Microsoft's artificial intelligence chatbot, Tay, which was designed to interact with users in the style of a nineteen-year-old woman. Less than a day after Microsoft launched the bot on Twitter, the combination of the bot's algorithms and the input it received from a particularly ruthless Internet apparently turned Tay into a racist. Among its tweets, Tay agreed with a white supremacist slogan, expressed support for genocide, and noted that Ricky Gervais "learned totalitarianism from adolf hitler, the inventor of atheism." Clearly, Tay's turn to bigotry was not meant to happen. Through this failure, the designers at Microsoft became better aware of how their program could interact with the raw, unfiltered id of the Internet and result in such hateful output, a type of output that they likely were not even aware was possible.

But waiting for unexpected behavior to reveal bugs is not enough. Many developers of technology actively search out bugs and collect them, placing them in a database so they can be addressed in a systematic way.

What's more, while software is in development, people try to actively break it, testing all the edge cases and weird things that users might do in real life, as opposed to the pristine manner in which programmers envision their creation being used. At Netflix, this strategy has even been taken to its logical conclusion in a piece of software known as Chaos Monkey. Chaos Monkey's function is simple: it unexpectedly takes Netflix systems out of service. Only by seeing how the vast Netflix system responds to these intentional failures can its engineers make it robust enough to withstand the unexpectedness that the messy real world might throw at it. The hope is that once Chaos Monkey has done its job there will no longer be any mismatch between how the engineers thought the system worked and how it actually does work.

Learning from bugs is an important mechanism for understanding any complex system. If we look at the history of science, we see that naturalists have been taking this approach to studying the complex systems of nature for centuries.

Naturalists for Technology

When I was young, the most incredible word for me—it looked so improbable it astonished me with its very existence—was "miscellaneous." This word was fascinating. It looked like it was cobbled together from so many different linguistic bits. I didn't know how to pronounce it, but it was wonderful nonetheless.

The magic of the word is not just in its sound and appearance. The cluttered spelling betrays its meaning: that there is a place in life for the grab bag. "Miscellaneous" means that even the messy and disordered is a category, a way of being organized. The existence of the miscellaneous is an affirmation that messiness—sprawling and complicated though it may be—can be tolerated and embraced.

Being comfortable with the miscellaneous—embracing a spirit of miscellany—is something that not everyone is good at. When we look at a complex situation, many of us, myself included, have a first instinct to somehow simplify it, to brush away all the complication and find the underlying elegance. When this works it can be incredibly satisfying, such as when we find a single cause for a failure. But when it doesn't work and we are left with a muddle, that's when many people become overwhelmed and unable to respond productively.

Naturalists who examine the natural history around us

have long been comfortable with the miscellaneous. Sometimes they detect an order in the living habits and behavior of the animals and plants they are observing, but there is also a purpose to their observations even without some sort of theoretical order—it allows them to understand and record the details, even if they don't yet have a complete mental framework for every living thing they see. The physicist Enrico Fermi, when asked to name one of the many particles studied in particle physics, replied, "Young man, if I could remember the names of these particles, I would have been a botanist." Naturalists—like John James Audubon, who, among other things, chronicled and illustrated the birds of the United States—recognize that it is important to know the details, and sometimes even the names, of these different individual pieces, whether or not we know how they fit together.

In the natural world, it is the bugs and glitches in biology, from mutations to diseases, that can teach us how living systems work. Errors in gene replication, from large aberrations in chromosomes to a single incorrect letter in DNA, and the visible differences or defects they produce, are how we can learn about the functions of genes. Mutations in fruit flies have helped to provide insight into how organisms develop from a single cell and a genetic blueprint into full-grown living things. For example, one way that biologists learned about key genetic sequences that govern body form was through the monstrous

Antennapedia mutants: flies that have legs sprouting from their heads in place of antennae.

In technology, we need the same sort of approach. The digital universe, to borrow a term from the historian of technology George Dyson, is expanding beyond our control. According to Dyson, there were only 53 kilobytes of high-speed RAM on the entire planet in March 1953. A single personal computer can now have more than a hundred thousand times this much RAM. The digital universe has become unimaginably richer, larger, and more interconnected. And it has become increasingly independent of humanity. Messages speed their way around the globe faster than we can recognize them, intersecting and interacting in often wonderfully unexpected ways.

While we cannot understand all the interrelationships of these systems, we can act as technological naturalists, chronicling and cataloging the diversity of the systems and parts of systems that we encounter. We can examine the anomalies and malfunctions to gain insights, even if we don't fully understand the system as a whole.

Chapter 5

THE NEED FOR BIOLOGICAL THINKING

In the mid-seventeenth century, an English physician named Nathanael Fairfax published some articles in the scientific journal *Philosophical Transactions,* the publication of the Royal Society, the British national scientific society. He observed several intriguing phenomena, and decided to communicate them to the scientists of the age.

In one paper, on "Divers Instances of Peculiarities of Nature, Both in Men and Brutes," Fairfax told the story of a man of about forty years old who was accustomed to drinking hot beer. When he eventually drank a cup of cold beer, he became sick and died within a couple of days. This observation led Fairfax to speculate about the agreeability of certain temperatures to the stomach. Fairfax also wrote of a woman who was struck with a sort of nausea every time she heard thunder. He did not speculate about why thunder might have this effect on a woman, noting only, "And thus it hath been with this Gentlewoman from a Girl."

One gets the sense that while Fairfax wanted to eventually learn something from these observations and facts he recorded, the act of making the observations was itself enough, at least as a first step toward understanding.

Around this same time, a young physicist named Isaac Newton was thinking about how objects move and how light works. While Newton was studying at Trinity College, Cambridge, a plague began to sweep through the country. As a precaution, the university closed. So Newton spent the next couple of years primarily at home, back in Woolsthorpe in the countryside. During this time, he made fundamental advances in calculus, optics, and understanding the motions of the planets. He conducted mathematical work, performed experiments such as poking himself in the eye socket to understand the nature of color, and even apocryphally observed the apple falling from the tree. Like Fairfax, Newton cataloged observations, but he also uncovered a set of universal principles, often mathematically described, that govern our physical world.

In a way, the work of these two men, occurring in the same country during the same time period, embodies two competing approaches to understanding the universe's complexity.

Newton's approach seeks to *unify* all the different things

that we see around us, simplifying this variety and diversity through a set of elegant explanations, and often a small set of equations or principles. We see this in Newton's formulation of the law of gravitation. In a single equation we find insight into falling objects, the ebb and flow of the tides, and the motions of the planets. The hope of such unity—a desire to discover an order underlying every aspect of the universe that we are aware of, and to place each component and particle in its place—is what drives physicists to search for a unified Theory of Everything. It's revealed as well in the tone of the scientist Thomas Henry Huxley, who felt that the "great tragedy of Science" is "the slaying of a beautiful hypothesis by an ugly fact." An elegant theory is the goal, and it is a tragedy of the highest level when something is found that contradicts it, or complicates it.

Fairfax's approach, on the other hand, eschews the pursuit of elegance to embrace *diversity* and complexity. This tendency accepts a certain messiness to the world, and celebrates learning new details, even if they are hard to immediately fit into a single framework. This approach is easily ascribed to, say, the butterfly collector, who catalogs and describes the many butterflies he discovers. We also find here the modern physician, the intellectual descendant of Fairfax, who marvels at level after level of the workings of the human body, from the complicated steps involved in our blood-clotting process to the intricate nature of enzyme cascades; or the astronomer who immerses herself in the many types and categories of galaxies revealed by powerful space telescopes.

The naturalist recoils at Huxley's complaint, for there is no such thing as an ugly fact. All facts and bits of knowledge provide new information on the wondrous complexity and diversity of the world around us. Rather than being upset when facts fail to conform to our mental models, we can delight in the unexpected, and find a new way to explain such surprising developments.

The physicist Freeman Dyson has described the Newtonian approach as the science of classical Athens, noting that this mind-set "emphasizes ideas and theories . . . [and] tries to find unifying concepts which tie the universe together." The diversifying approach, he writes, can be described as that of Industrial Revolution–age Manchester: it "emphasizes facts and things; it tries to explore and extend our knowledge of nature's diversity."

Dyson further notes that there is an additional way of identifying these two perspectives, with biology as the domain of the diversifiers and physics as that of the unifiers. I term these two perspectives *physics thinking* and *biological thinking*. Within physics there is a distinct trend toward unifying and simplifying the phenomena observed. This is embodied by the work of Einstein or Newton or James Clerk Maxwell, who developed a handful of equations to explain the workings of electricity and magnetism. Simplification, even oversimplification, is revered within the realm of physics.

On the other hand, biologists, as a rule, have a greater comfort with diversity and bundles of facts, even if they are left unexplained by any single sweeping theory. A smaller, more qualified and modest model is just fine. Of course, this is not always true, as Charles Darwin was clearly a unifying force within biology, and many molecular biologists, applied mathematicians who specialize in mathematical biology, and many other types of biologists also tend toward this unifying approach.

In the end, both of these traditions seek the development of theories that are general and predictive. However, the two modes of thinking go about this in different ways, and their differences, driven by the properties and relative complexity of the systems they study, can be examined through their relative comfort with abstraction. For example, the use of mathematics to abstract away details at a grand level is found everywhere in physics, but less often in biology.

This is clear from the following version of an old scientific joke. A dairy farmer, interested in increasing the milk yield of his cattle, brings in two consultants to help him: a biologist and a physicist. The biologist goes off and after a week comes back with a detailed report on what to do for each cow, depending on weather conditions, the size and type of the cow, and so forth. The report is over 300 pages long, but the farmer is assured that following the various procedures will result in an average increase in milk yields of about 3–5 percent. The physicist goes off and comes back three hours later, announcing a general and powerful solution to increase yields by over 50 percent. How so? asks the farmer. "Well," replies the physicist, "first, you assume a spherical cow. . . ."

Abstraction has its place, but it is not in assuming spherical cows. When details are abstracted away in biology, not only is information lost, but you often end up losing significant portions of what the world contains and fail to explain what's important, such as the edge cases.

Biological thinking and physics thinking are distinct, and often complementary, approaches to the world, and ones that are appropriate for different kinds of systems.

The Kind of Thinking That Technology Requires

How should we think about complex technologies? Are they biological systems, or physics systems? Which mode of thinking does technology require? It's time to explore the characteristics of each type of system and compare them to what we know about technology.

First, biological systems are generally more complicated

than those in physics. In physics, the components are often identical—think of a system of nothing but gas particles, for example, or a single monolithic material, like a diamond. Beyond that, the types of interactions can often be uniform throughout an entire system, such as satellites orbiting a planet.

Not so with biology. In biology, there are a huge number of types of components, such as the diversity of proteins in a cell or the distinct types of tissues within a single creature; when studying, say, the mating behavior of blue whales, marine biologists may have to consider everything from their DNA to the temperature of the oceans. Not only is each component in a biological system distinctive, but it is also a lot harder to disentangle from the whole. For example, you can look at the nucleus of an amoeba and try to understand it on its own, but you generally need the rest of the organism to have a sense of how the nucleus fits into the operation of the amoeba, how it provides the core genetic information involved in the many functions of the entire cell. As our technologies become more complex and intertwined, it's clear that they resemble biological systems more than those of physics.

Second, biological systems are distinct from many physical systems in that they have a history. Living things evolve over time. While the objects of physics clearly do not emerge from thin air—astrophysicists even talk about the evolution of stars—biological systems are especially subject to evolutionary pressures; in fact, that is one of their defining features. The complicated structures of biology have the forms they do because of these complex historical paths, ones that have been affected by numerous factors over huge amounts of time. And often, because of the complex forms of living things, where any small change can create unexpected effects, the changes that have happened over time have been through *tinkering:* modifying a system in small ways to adapt to a new environment.

For example, many of the most important sequences of DNA in a human cell, such as the ones that power how our genetic code is translated or how we use energy, are the same ones that other, far different organisms—separated from us by eons—also use. This biological legacy code sometimes remains unmodified, but often, through evolutionary time, these systems are also tinkered with and changed.

On the macro scale of an organism, this means that new functions are often layered on top of old ones, which can sometimes lead to problems. For example, we may walk on two feet now, but our skeletal structure initially evolved for more quadrupedal locomotion, that is, on all four legs. Evolution tinkered with the bodies of our ancestors, giving us a "good enough" means of walking with an upright spine. But the solution is far from perfect, and many representatives of our species suffer from back pain.

What is the result of this phenomenal complexity and dependence on the path taken over the course of evolution?

When it comes to understanding these systems, biologists cannot live by aesthetics. As the biologist Steven Benner notes, an evolved system works, but it needn't be beautiful. It can be utilitarian, which we sometimes confuse with beautiful, but it can be far from elegant or simple. Benner even notes that the structure of DNA is not really beautiful, just rendered so by artists through the use of abstraction. Biological systems are generally "hacks" that have evolved to be good enough, rather than pretty, designed systems. They are kluges.

Evolution can even leave us with obsolete legacy code, just as technology does. My backyard in the Midwest is home to a number of honey locust trees, which sport an array of large, dangerous-looking thorns. These thorns seem to have no purpose other than making me worried that I will inadvertently impale myself upon them. So why do they exist? One theory is that they are there to protect the leaves of the tree from being eaten by a mammoth, a giant ground sloth, or another species of North American megafauna. Except, of course, these creatures are long extinct. The information that codes for the honey locust's seed pods seems also to have evolved under the evolutionary pressure of those same now-extinct megafauna, yet it still exists.

Similarly, while there is still a great deal of debate on this matter, some scientists argue that there is a lot of extra material in genomes that seems to have accumulated and persisted, despite lacking any particular biological function (what is sometimes referred to as "junk DNA"). Just as many complex technological systems, including software, contain features that no one uses and that might even be completely obsolete, many biological systems also have vestigial features whose original functions are no longer relevant.

Of course, the parallels between biology and technology are not perfect. Biology handles legacy code differently: the honey locust might eventually lose its thorns. If this trait is truly useless, then producing thorns is a wasteful expenditure of energy for the honey locust. Over evolutionary time, the thorn trait will be swept away by the success of variants of the tree that lack thorns and are therefore "fitter." My descendants, freed from the risk of being impaled by this tree, will be thankful. No such parallel exists with most of our technologies: software programs don't automatically sweep away their own legacy code because it's outdated and inefficient.

Finally, the similarities between biology and technology can be seen in the concept of highly optimized tolerance, mentioned in the previous chapter. Technologies can appear robust until they are confronted with some minor disturbance, causing a catastrophe. The same thing can happen to living things. For example, humans can adapt incredibly well to a large array of environments, but a tiny change in a person's genome can cause dwarfism, and two copies of that mutation invariably cause death. We are of a different scale and material from a particle accelerator or a computer network, and yet these systems have profound similarities in their complexity and fragility.

Overall, there is a deep kinship between biology and technology—which means there is something to be learned from how biologists think.

Field Biologists for Technology

As our technological systems become more complicated, we often are left with only the extremes of understanding: either a general notion of how the thing works, even if its innards are at best murky to us, or an examination of its bits and pieces, without an inkling of how it all fits together and how we can expect it to behave. The first is a tendency toward physics thinking, while the latter leans toward biological thinking.

In the face of increasing complexity, some choose to rely on the physics approach, abstracting away details to get a general sense of the system. For example, when looking at a complex social system, such as a company or a city, a physics-style approach for making sense of it might be to graph one of its properties and see if it conforms to a specific mathematical curve. This can yield some insight—or at least a hint of what is going on—but when there are so many different reasons that this system might fit that curve, you can be left with more questions than answers. These systems are often not amenable to immediate and large-scale abstraction; they are too messy and too complicated.

Biological thinking, therefore, with its focus on details and diversity, is a critical perspective for dealing with a messy evolved system that can be completely understood only through a lot of initial prodding and testing. The way biologists, particularly field biologists, study the massively complex diversity of organisms, taking into account their evolutionary trajectories, is therefore particularly appropriate for understanding our technologies. Field biologists often act as naturalists—collecting, recording, and cataloging what they find around them—but even more than that, when confronted with an enormously complex ecosystem, they don't immediately try to understand it all in its totality. Instead, they recognize that they can study only a tiny part of such a system at a time, even if imperfectly. They'll look at the interactions of a handful of species, for example, rather than examine the complete web of species within a single region. Field biologists are supremely aware of the assumptions they are making, and know they are looking at only a sliver of the complexity around them at any one moment.

Similarly, when encountering a complicated tangle of a technological system, whether a piece of software, a country's laws, or the entirety of the Internet, a physics mind-set will take us only so far if we try to impose our sense of elegance or simplicity upon its entirety. If we want to understand our technological systems and predict their behavior, we need to become field biologists for technology.

What does this mean? When we're dealing with different interacting levels of a system, seemingly minor details can rise to the top and become important to the system as a whole. We need "field biologists" to catalog and study details and portions of our complex systems, including their failures and bugs. This kind of biological thinking not only leads to new insights, but might also be the primary way forward in a world of increasingly interconnected and incomprehensible technologies.

As discussed in the previous chapter, we can learn from bugs in technological systems, just as biologists learn from genetic errors. But biologists do much more than simply learn from these glitches. To better understand how to think like a biologist, we must look at how they conduct their work more generally.

One of the major advances of recent years in genetics is RNA interference, or RNAi. This involves using short pieces of RNA (a "cousin" of DNA that our cells produce and use for myriad purposes) to deactivate the production of proteins. By constructing the right RNA "text," you can effectively switch off certain genes.

How was this mechanism discovered? One of the initial steps in the discovery involved an attempt by some geneticists at a biotech start-up to make a more purple petunia. They had identified the gene responsible for making the purple in the petunia, and thought that adding another copy of the gene would yield a flower with a richer hue. Instead, when they attempted this, they got a white flower—the opposite of what they'd expected. The inserted genetic information ended up making no purple color instead of making more of it. Rather than sweep the unexpected outcome under the rug as something to be ignored, the researchers noted it for further examination, and it eventually led to the development of RNAi.

This kind of thing happens all the time in science, biology or otherwise. Isaac Asimov is reputed to have noted the following: "The most exciting phrase to hear in science, the one that heralds new discoveries, is not 'Eureka' but 'That's funny . . .'" Penicillin was discovered when Alexander Fleming saw something odd on a petri dish. The atomic nucleus was discovered when scientists noticed unexpected results in an experiment that involved shooting radioactive particles at a thin piece of gold foil. The scientists could have discarded these results. Instead, they looked more closely—and saw the basis for an entirely new understanding of atomic structure. By cultivating this natural history mind-set of cataloging and collecting the bits and pieces that don't make sense, we can learn new things about whatever we're studying.

But simply waiting to observe the unexpected may not be enough. Biologists will often be proactive, and inject the unexpected into a system to see how it reacts. For example, when biologists are trying to grow a specific type of bacteria, such as a variant that might produce a particular chemical, they will resort to a process known as mutagenesis. Mutagenesis is what it sounds like: actively trying to generate mutations, for example by irradiating the organisms or exposing them to toxic chemicals.

While this sounds harsh, it has a purpose. When a system is so complex that it is hard to anticipate how, exactly, it might respond—and what changes in a genome might yield the desired effect—one often needs to use a certain amount of randomness to find out what the system can do. Essentially, these systems are so highly nonlinear and complicated that we must actively use an evolutionary process of tinkering in order to discover how they work.

We can see a similar principle at work in the search for new medicines. In pharmaceutical research, one way that new drugs are created is through an active process of sifting and checking countless variations of chemicals in order to find one that provides the desired effect. The actual chemical mechanism might be only imperfectly understood, but the testing process—the discovery of the rare and effective—provides a way of understanding the human body, even if only dimly and indirectly at times. This poking and prodding of the system in order to learn more and to sift out molecules that can be effective pharmaceuticals is another tinkering approach to discovery.

Applying biological thinking to technology involves recognizing that tinkering is a way of both building a system and learning about it. As Stewart Brand noted about legacy systems, "Teasing a new function out of a legacy system is not done by command but by conducting a series of cautious experiments that with luck might converge toward the desired outcome." Such is the approach of a "field biologist for technology." We saw this put into practice in the previous chapter with Netflix's Chaos Monkey, which is essentially a mutagenic piece of software designed to introduce errors in order to learn about the system and improve its reliability. Glitches, even ones introduced by us, must be chronicled and examined in hopes of gaining a better sense of how the greater system operates.

This biological approach can also aid us in understanding disasters and catastrophes—much the way biologists think about cancer. When cells begin to grow into a tumor, it is rarely just a single thing that has gone wrong. Rather, cancer can result from an accumulation of many factors and biological responses that interact in complex ways, leading to a large-scale failure: a potentially deadly disease. Recognizing that such an accumulation of problems and responses can cause similar failure cascades in our technologies means thinking more like a biologist. For example, in a nuclear power plant, small problems can add up and cause serious issues—multiple independent causes seem to be implicated in what eventually led to a partial meltdown at the Three Mile Island plant. We can also identify interactions that cause massive problems in the world of finance.

Happily, we are beginning to get help in these endeavors

from our technologies themselves. There are now computational tools that can help find unexpected outcomes in a system, part of the domain known as "novelty detection." Machines might be able to act in partnership with these field biologists and naturalists, helping us to better understand—even if only partly—our own technologies.

When Physics and Biology Meet

The biological aspects of technology—its klugeyness, its growth and change due to evolutionary tinkering, its many miscellaneous details—are extensive. But does this mean that we should abandon our search for underlying regularities in all this complexity? Absolutely not. Physics thinking still has a role in how we approach technology.

When attempting to understand a complex system, we must determine the proper resolution, or level of detail, at which to look at it. How fine-grained a level of detail are we focusing on? Do we focus on the individual enzyme molecules in a cell of a large organism, or do we focus on the organs and blood vessels? Do we focus on the binary signals winging their way through circuitry, or do we examine the overall shape and function of a computer program? At a larger scale, do we look at the general properties of a computer network, and ignore the individual machines and decisions that make up this structure?

These are not always easy questions to answer. Sometimes we must tend toward physics thinking, abstracting away the details to understand the system as a whole. And sometimes the details are important, as with our hapax legomena and edge cases: then we must rely on more biological thinking.

But all too often, the different levels of resolution collide. Sickle-cell anemia, a quite serious systemic disease, is caused by a tiny change in a single base pair in our DNA. A large fraction of the United States electrical grid can be brought down by a cascade set off by trees touching power lines in Ohio, as happened in the summer of 2003. When systems become more and more interconnected, not only do resolution levels intersect, but domains thought to be separated are increasingly brought together. More and more we need to combine both the physics and the biological ways of thinking, looking at the order while not ignoring the rough edges. A biological mind-set partnered with a physics mind-set allows us to feel more comfortable with the kluges around us. In Neal Stephenson's novel *Cryptonomicon,* one of the characters elaborates on the structure of the pantheon of Greek gods, making exactly this point:

> And yet there is something about the motley asymmetry of this pantheon that makes it more credible. Like the Periodic Table of the Elements or the family tree of the elementary particles, or just about any anatomical structure that you might pull up out of a cadaver, it has enough

> of a pattern to give our minds something to work on and yet an irregularity that indicates some kind of organic provenance—you have a sun god and a moon goddess, for example, which is all clean and symmetrical, and yet over here is Hera, who has no role whatsoever except to be a literal bitch goddess, and then there is Dionysus who isn't even fully a god—he's half human—but gets to be in the Pantheon anyway and sit on Olympus with the Gods, as if you went to the Supreme Court and found Bozo the Clown planted among the justices.

The more we examine the systems around us with open eyes, we see this balance between biology and physics. We find it in our ecosystems and in the chaos of technology that we rely on every day. We find it in the Greek pantheon, and in many other stories we tell ourselves.

Storytelling, in fact, allows us to indulge our desires for either biological or physics thinking. Some stories are finely crafted machines with no extraneous parts; everything fits together. We see this in "Chekhov's Gun," dramatist Anton Chekhov's principle that any element introduced in a story must be crucial to advancing the plot. A loaded rifle introduced early in the first act of a play must go off by the third.

On the other hand, there are some stories in which color is added, creating a richness of experience without necessarily moving the plot along. Homer's catalog of the invading

Greek ships in the *Iliad,* and Kramer's growing list of never-seen eccentric friends on *Seinfeld*—Bob Sacamano, Lomez, Corky Ramirez—are not essential plot points, but they are important. They are the biology alongside the physics: both are needed to create the rich world we inhabit when we engage with a story.

In the field of special effects, there is a delightfully evocative term: "greeblies." When I hear it, I think of gremlins and gibberish mutterings. Greeblies are the little bits and pieces that get added to a scene, or to a single object, to make it look more believable. You can't have a futuristic starship that is all angles and smooth sides; you need to add ports and vents and sundry other impenetrable doodads and whatsits, pipes and bumps, indentations and grooves. Think of the ships in *Battlestar Galactica* or *Star Wars.* They are more visually intriguing thanks to their complications of unknown purpose.

This process of greebling is closely related to a well-known quote from the mathematician Benoit Mandelbrot, who coined the term "fractal": "Why is geometry often described as 'cold' and 'dry'? One reason lies in its inability to describe the shape of a cloud, a mountain, a coastline, or a tree. Clouds are not spheres, mountains are not cones, coastlines are not circles, and bark is not smooth, nor does lightning travel in a straight line."

So, too, our technological systems, once embedded in the real world, are far from the cleanly pristine logical constructions of the drawing board; they are full of the miscellaneous details of biology that have accreted over time, much like the evolutionary hodgepodge found within living systems. Our stories are built with details and complications, and so are our complex systems. In fact, we seem to recognize something as more "realistic" when it is complicated, full of tiny crenellations and details that often elude our understanding. Ultimately, we need this messiness, we need the greeblies, even if at another resolution we might abstract it all away.

Biological thinking needs to exist alongside physics-based thinking. Recall "Funes the Memorious," the short story by Jorge Luis Borges where the title character is burdened by a perfect memory. While many of us would view this as a gift (as does Funes), it's not quite that. The reader learns that for Funes, nearly every detail and perspective generates a new memory and in turn a new category, a new kind. When gazing at a dog, Funes doesn't just see a dog; he sees a specific dog from one angle, and then, as soon as it or he moves, a completely new dog memory is created. Funes does not unify the particulars into general concepts because his memory is too detailed. He can no longer form any abstractions because his tolerance for complications is too great.

In fact, the end goal of biologists is to create models and identify regularities, even if on a smaller scale. So, when confronted with a complex piece of technology, we must begin by acting like field biologists, experimenting around its edges to see how it behaves, with the end goal of some degree of generalization. This is actually how a lot of people approach open-ended video games like *Minecraft.* You first collect huge amounts of information about your virtual world—what you can do, what you can't, what kills you, how you successfully survive—and then begin to make little mental models, small-scale generalizations within a much larger whole.

Or, when you are working with an advanced piece of software such as a gargantuan word-processing tool, and the endnotes in your document go haywire, do not panic. Instead, look at what went wrong: Did several endnotes all have the same numbering? Do they still connect to the correct places in the text? And so on. By being willing to tinker—a cue we take from living things—you get a better sense of the details and nuances of a very complex system. Acting as field biologists for technology allows us to look for and closely study the various pieces of our constructed world, while still recognizing that they are only a tiny part of a much bigger and highly connected whole.

We now turn to one field that might help us accomplish this delicate balance in how we grasp technological systems.

The Science of Complexity

One natural path for managing and understanding complex systems is through complexity science: the quantitative study of these vast and complicated interconnected systems, ranging from living things and ecologies to the World Wide Web and even collaborations between film actors (think "Six Degrees of Kevin Bacon," where you try to connect every actor back to Bacon through film co-appearances). This is a rich and exciting field—one that I am a part of—and it uses a variety of powerful ideas and mathematical frameworks in a quest to find patterns and meaning in these complex systems. These approaches range from understanding network structures to developing computer models with huge numbers of interacting entities, known as agent-based models.

One of the main ways of grasping these systems, however, is by abstracting away a certain amount of messiness in order to find regularities that are amenable to clear mathematical shapes and understanding. For example, if you look at a massive network of interactions—such as who follows whom on Twitter—you can ignore the details of the interactions and note that the number of connections between individuals follows a specific category of probability curve known as a heavy-tailed distribution. This curve makes no claims about who is connected to whom or who has a lot of connections and who has few, but it shows that there is an underlying statistical regularity.

Other models in complexity science explore what is known as percolation: how something diffuses based on the structure of some sort of porous space, such as how petroleum moves through a rock. These models are very powerful at showing how small alterations in the density of the material can yield huge shifts in the ability of the fluid to move through the system, though they again make no claims about the actual specific locations of the pores, for instance.

Each model in complexity science is very good at providing a different angle of insight into what we are trying to understand. In fact, in many cases we can find simple mathematical models that seem to help unify whole swaths of our world. For example, there is a type of model known as a *diffusion-limited aggregation.* While on the surface this model looks like nothing more than an abstract computational plaything, it can be viewed as a unifying pattern for a wide variety of real-world phenomena.

The diffusion-limited aggregation, or DLA, begins by placing a tiny dot in the center of a giant grid. You then randomly walk other dots around the grid until one of them hits the original dot, and then it stops. That dot is now part of the aggregation. The dots that are still meandering randomly continue to do their walk until they hit the slowly growing DLA, and stop as well. Slowly, you keep on adding moving dots to this growing aggregation.

As you keep adding new randomly jittering dots to this thing, you end up getting a weird shape. It's not a blob, it's not a misshapen circle, or even a regular crystal. It's something far more organic. It looks like a spindly growth, and in fact, depending on the specific type of model, DLAs can look like coral, bolts of lightning, even cities. A simple, almost trivial computer model can create rich pictures.

This model and ones like it are almost paradoxical, because they seem to imply that massively complicated systems can be reproduced by really simple models. Of course, the details are not the same. These little toy models have a sort of "Potemkin complexity"—they look complex but contain an underlying simplicity, as opposed to a real giant city with the huge number of decisions that have gone into how it currently looks, or even a coral with all the details that went into how it grew. But sometimes these models are enough; for some purposes—such as trying to understand an entity's basic organization—the overall shape is all we care about. This abstract and simplified approach to complex systems can help us understand the shape of a system's behavior. It can teach us about feedback, interconnectivity, and the profound dependence of such systems on initial conditions. It can even place boundaries on our expectations for how a system might respond to changes. Being able to think effectively about complex systems using the concepts of complexity science is a skill necessary for anyone, expert or not, who wishes to engage with our ever-more complicated world. It is increasingly important to educate people as to how complex systems work, making these properties more intuitive.

Those aspects of complexity science that generalize have

their limits, however. When it comes to understanding the particularities, simplifying models are not sufficient. As the science writer Philip Ball has noted, "The patterns of a river network and of a retinal nerve are both the same and utterly different. It is not enough to call them both fractal, or even to calculate a fractal dimension. To explain a river network fully, we must take into account the complicated realities of sediment transport, of changing meteorological conditions, of the specific vagaries of the underlying bedrock geology—things that have nothing to do with nerve cells."

Happily, the tools of complexity science are not only used to create simplifying abstractions. They can also provide a window into the details of a system by actively sifting through its complexity in a rigorous fashion. For example, a team of researchers analyzed the United States Code using approaches derived from software engineering and complexity science, in order to determine various features of this body of law. Some of the analyses did provide results just about the overall features of these laws, such as the varying levels of complexity of different sections, or that there is a certain profile of complication in the Code. But their analyses also highlighted portions of the law that are particularly interesting or exceedingly byzantine. For example, the section detailing the "Powers and duties of the corporation" popped out as especially complex within the portion of the Code that deals with banking, at least in terms of the number of conditionals (if-then statements, just like in software). In the section dealing with taxes, the subsection on the "Qualified pension, profit-sharing, and stock bonus plan" also had a high complexity. These kinds of results can help us zoom in and focus on the parts that are most complex, and perhaps even take stabs at simplifying them.

In the same vein as field biology, certain complexity science approaches can be used to learn about the behavior of a subset of a complex system rather than the system's behavior as a whole. Approaches can even be used to find the outliers: the parts of the complex system that don't fit the generalized principles. Ultimately, all these approaches can help us learn which parts of a system it might be worthwhile to look at more closely, in hopes of a better understanding of how they all interact. We must balance the complexity science that abstracts away information with the type of analysis that finds the particulars that don't fit neatly into the model.

We again are left with the tension between thinking informed by physics and by biology. On the one hand, we yearn for a simple elegance in our technological world, and wish the convoluted away. But on the other, acknowledging something as complicated—particularly something that has grown and evolved over time—is a sign of nuance, a sign of maturity. As we mature as individuals, we recognize complications in our relationships, nuances in our interactions with others. As we have matured as a society, we must recognize the complications and irregularities inherent in our constructions. Complexity science, while far from a panacea, can help us strike this balance: highlight details to focus on, but also place boundaries on our knowledge and our level of concern. For example, complexity science can show us how easily systems can become unstable, and where we must direct our efforts and attention. If a simple model demonstrates that a large technological network can be wiped out by only small changes, we can no longer remain blissfully ignorant of this fact.

That being said, finding the right balance between physics thinking and biological thinking is not easy. It's a difficult process, and one that has long played out in the history of ideas.

Making Sense of the World: A Very Short History

The ancient world was filled with individuals trying to make sense of their surroundings. One of the first places where this kind of thinking arose in a rigorous fashion was Ionia. Ionia was located on the Aegean Sea, on the western coast of modern Turkey, and was a collection of cities where the first philosophers of the Greek world, even before Socrates, began to ruminate on the cosmos. The ideas of these philosophers, part of the group often known as the pre-Socratics, were almost uniformly ones of wide-sweeping simplicity and generalization.

Xenophanes, an Ionian poet-philosopher, argued one of two things, depending on what you read: that everything seemed to arise from only two sources, earth and water, or that all derived from a single source, earth. Either way, Xenophanes sought a unifying principle or two to explain the complexity of the world.

On the other hand, Thales, another Ionian, held that all was water. Another philosopher thought everything was related to fire. Captain Planet would have liked the Ionians.

Somewhat differently, Anaximander of Miletus, yet another Ionian, felt that everything came from a single infinite first principle. He also had numerous other precise concepts, such as that animals came from moisture "evaporated by the sun," or that the Earth was surrounded by a circle twenty-eight times its size, from which he derived astronomical insights.

But overall, many of these philosophers held to a small handful of general principles and fundamental materials that could explain the world. More generally, they seemed to adhere to the idea of the Greek *kosmos*. Unlike the English word "cosmos," the Greeks didn't just use this term to refer to the universe and the totality of all things. Bound up in this term is the elegance and beauty and ultimate order of the universe. The elegance of nature—and the physics mode of imposing order—is therefore implicit in this term for nature itself. These philosophers also used the term *arche,* which might be most clearly associated with a guiding rule or first principle that underlies the universe. It was a unifying wish, a wish for order around us.

Jumping forward to the dawn of the modern era, we see a bit of a shift, to an increased focus on the bizarre and the unexplained. We see the miscellaneous. According to Philip Ball, anomalies and eccentricities were relatively uninteresting to the ancients and to medieval scholars because they were interested in affirming what was already known. But for those building wunderkammers on the cusp of the modern era, the anomalous and weird and strange were precisely the most exciting parts of the world. The historian of science Lorraine Daston has argued that there was a strong penchant in the early days of science—far beyond Nathanael Fairfax—for delighting in "strange facts."

From "Observables upon a Monstrous Head" to "On a Species of Wild Boar That Has a Hole in the Middle of Its Back, Which Foams When It Is Pursued by Hunters" or "A Narrative of Divers Odd Effects of a Dreadful Thunderclap"—all article titles from the seventeenth century quoted by Daston—we glimpse bizarre bits of information, and the scientists of the day who delighted in them. There was a desire to chronicle and ponder the unexpected and the weirdest aspects of the natural world. As Daston notes, "The first scientific facts were stubborn not because they were robust, resisting all attempts to sweep them under the rug, but rather because they were outlandish, resisting all attempts to subsume them under theory." We don't often think of science nowadays as a means of collecting the strangest things we can find—with no way as yet of explaining them—but that is what science was, and it still is in many ways. We learn the most when we try to actually confront the aberrations and exceptions around us, such as the physical oddities that led to discovering the nature of the atomic nucleus or the phenomenon of RNAi.

This is the kind of biological thinking, tempered with physics thinking, that must be imported into technology. But while we can all participate in this endeavor to some extent, whose job is it to do this technological fieldwork, to cultivate the discovery of the unexpected in a complex technological edifice and foster a sense of play with the incomprehensible? In our age of specialization, it perhaps becomes the role of the generalist to recognize these details, the unexplained edge cases and rough joints of our systems.

The Return of the Generalist

In 2009, a team of scientists examined hundreds of millions of interactions with online scientific papers in order to discern the "clickstream" of readers, the path they take from one publication to the next. This data revealed patterns of how people moved from one subject area to another, even generating a beautiful image of connections between areas. For example, in their map, nursing is near psychology and education. Organic chemistry bridges physical chemistry and analytic chemistry, economics is connected to sociology and law, and the field of music stands somewhat distinct.

Of course, these are oversimplifications. Music actually incorporates concepts from physics and psychology, while economics draws heavily from mathematics. But examining this clickstream is one way to explore the interconnected nature of ideas. Even though we continue to specialize in order to handle the more complicated systems we are building, seeing this web of interconnections reminds us that each domain does not stand alone; they are all part of a vast connected framework.

Since these systems are interconnected in many different ways, we will increasingly require the ability to connect one area of knowledge to another. When constructing a computer program that can play *Jeopardy!*, for example, you need knowledge of everything from linguistics to computer hardware; specialization alone will not work. We need a certain breadth of knowledge. However, as noted earlier, before too long, we will bump up against the limits to what we can truly understand; we just can't hold all the relevant knowledge in our heads.

In response, we need to cultivate generalists, individuals

who not only can see the lay of the land—the abstract physics style of thinking—but can also delight in the details of a system without necessarily understanding them all—the more miscellaneous biological style of thinking. Generalists might be the individuals most suited to acting as naturalists and field biologists for our constructed complex systems. They can jump from one piece to another, examining the parts that don't make sense and seeing hints of what is going on in these vast technologies.

But given the vast growth of knowledge, can we still have generalists?

It is still possible, but it's hard. Creating generalists who are able to serve this function well in our society first involves the construction of what have become known as T-shaped individuals, a term that appears to have first originated in computing education. T-shaped individuals have deep expertise in one area—the stem of the T shape—but breadth of knowledge as well: the bar of the T.

What do these types of people look like? One example is the data scientist, who uses the tools of computer science and statistics to find meaning in large datasets, no matter what the discipline. Data scientists have to know a lot about many different areas in order to do their job successfully. We see something similar in applied mathematicians, who use quantitative tools to cut across disciplines and find commonalities, acting as generalists.

Generalists can be found in such areas as consulting and book editing, and you can even find them in the world of venture capital, where there are many people who are knowledgeable in multiple different areas and can use this expertise widely. If someone wants to invest in outer space, 3-D printing, agricultural technology, tools for scientific discovery, and more, one had better be at least somewhat of a generalist.

These individuals embody both expertise and the polymath tendency to explore many different domains. But cultivating such T-shaped individuals, who combine specialization with some measure of generalism—people who can at least begin to handle some of the growing complexity around us—is not that simple.

Right now, the job market rewards specialization, making it difficult to educate T-shaped people. And I think that we will not find these types of people within academia for a long time to come. As business professor David Teece has noted, it was a lot easier to be a Renaissance man during the Renaissance. But one hint of how this change might occur can be seen by looking to the Girl Scouts.

The Girl Scouts once offered a fascinating kind of merit badge: the Dabbler badge. This allowed a young Scout who wanted to do a little bit of everything to not only generalize, but be recognized for that achievement. Don't want to specialize in ceramics only, or just in photography, but you think some degree of expertise in a whole lot of things could be useful? Then the Dabbler badge is for you. My mother recalled that when she was in the Girl Scouts, the Dabbler badge was the only one she earned.

Perhaps it's time for the knowledge equivalent of the Dabbler badge: a way to acknowledge and foster those dabbling in different ideas, all the way from grade school to late in one's career. These people can work in concert with the specialists, helping to translate from one field to another and stemming the tide of incomprehensibility. These dabblers can bring biological thinking to the massive systems we are building, searching for glitches that can guide our understanding, and cataloging the strange and amazing parts of these edifices.

In fact, it seems that the places where generalists can thrive best are the places where we understand the least, where the systems are so complicated and interconnected that the best we can do is hope for a chronicling of the miscellaneous. What this means is that the education of generalists will involve not just learning what is known, but also learning ways of exploring the unknown, the new, and the unexpected. This will involve rigorous thinking in the liberal arts, as well as such skills as computational thinking, logic, and even data visualization. Or take the talents of the title character of *Hild,* a novel set in the Early Middle Ages of Britain. Hild is recognized for her ability to weave tapestries as well as for her keen intelligence.

When others speak of her skill with weaving and the loom, they describe her as having a "pattern-making mind." Her intelligence is the same way. This pattern-making mind is able to construct connections and see a web of associations, helping Hild to navigate the geopolitical intrigue that surrounds her.

Of course, generalists alone are relatively useless. They are best when working alongside specialists, helping in the process of translation and communication, or playing other roles complementary to the still-important contributions of specialists. This can mean having a generalist working on each large project in a company or other organization, having a small team of roving generalists in a large corporation who can provide context and value for the entire company, or even having small teams of generalists-for-hire who work with specialized firms to supplement their focused expertise.

But a generalist is more than simply a T-shaped person, someone with a bit of a broader background than is usual these days. Becoming a generalist involves a willingness to consciously stitch together authentically different fields, even at the risk of failing to do so. The best generalists pair a chronicling of details with a pattern-making mind. A pattern-making mind needn't be one that thrives on abstraction and generalization alone, but one that makes connections and analogies. Seeing interrelationships and interactions is necessary in highly complex systems. The pattern-making mind can make sense of these systems, even if only partial sense, through a combination of intuition and biological thinking.

I'm a member of The Cloud Appreciation Society. I am of the opinion that a clear, burning blue sky—while beautiful—sometimes needs a bit of contrast in order to be properly appreciated. It needs clouds. When I first told my wife of my membership, she thought I was making up some bizarre organization, until the level of detail I provided—the lifetime membership, small pin, and official certificate—all mentioned with a straight face, made this seem increasingly unlikely. I really find clouds fascinating and beautiful. But I am little more than an enthusiastic amateur when it comes to understanding them.

To the untrained eye, the variety of clouds is broad but rather unremarkable. You've got your storm clouds, your fun fluffy ones, the streaky ones, and so on. But this relatively childish taxonomy conceals a great deal of additional complexity and variety. There are clouds named cumulus, cirrus, nimbostratus, cumulonimbus—some of the common kinds, using the basic descriptive roots—and the specific subtypes, such as fractus, shelf cloud, lenticularis. In fact, a scientific classification of clouds was developed only in the past couple of centuries. As the science bookseller John Ptak notes, "Clouds though pretty much escaped the notice of even the greatest of all great classifiers, Aristotle, and just about everyone else, [and there was,] for two dozen centuries, no real scientific approach to them until Luke Howard first published on his cloud classifications . . . in 1803."

Elsewhere in the sky, we find phenomena inspiring numerous other wonderful meteorological terms. From sheet lightning to St. Elmo's fire to ball lightning, these terms are many and weird. When it comes to weather science, we have made great strides in prediction and explanation. But throughout history we have been confronted with new and strange phenomena, which in turn have helped provide new insights into how our atmosphere goes about its business.

Clouds and other atmospheric phenomena aren't made by us, but how we approach them can give us a glimpse into the proper frame of mind for dealing with our complex technological systems. When we don't understand a phenomenon or a system, or find it frightening, we shouldn't ignore it. Even if something can't be explained, it still has a place, the category of the yet-to-be-understood, which can then be used as a wedge for further insight.

Furthermore, when we are confronted with something we can't fully understand, we can focus on the details in the system, understanding specific parts of the whole. Ultimately, a biological approach to a system is an iterative and tinkering-based one, guiding us toward further understanding by using details and unexpectedness to constantly gain further insight.

There is whimsy and beauty in the complicated and the

unexpected. A glittering and shimmering technological network, with its branching gossamer web of links and interactions, is unbelievable in its complexity. And sometimes, even if we don't understand every part and every whole, an imperfect grasp can be enough. We can walk humbly with our technology.

Chapter 6
WALKING HUMBLY WITH TECHNOLOGY

Moses Maimonides was a singular individual in the history of philosophical thought. Trained as a physician and rabbi in the twelfth century, he eventually served as a court physician in Egypt. He codified Jewish law, while at the same time innovating and incorporating contemporary scientific ideas. There is a saying within Judaism that "From Moses to Moses [Maimonides], none arose like Moses." Maimonides also wrote a philosophical text, a masterpiece called *The Guide of the Perplexed.* This book is a bold attempt to blend together Jewish thought with Aristotelian philosophy, which was considered some of the most advanced philosophical thinking at the time. *The Guide of the Perplexed* was read widely both within and outside the Jewish community.

What, then, can we learn from this scholar, a man whose

curiosity led him to examine everything from nutrition and astronomy to politics? He recognized that, despite the incredible human intellect, there was a fundamental mismatch between our curiosity and what we could actually understand, at least when it came to the infinite, which was necessarily beyond the human mind. As a man of his time, though in many ways ahead of it, he held that the only one who could actually understand such matters was God, though the God of Maimonides was a more philosophical and abstract concept than the Great Man in the Sky, the then-common conception of the divine. But Maimonides was certain of our mental limits: ". . . for man's intellect indubitably has a limit at which it stops. There are therefore things regarding which it has become clear to man that it is impossible to apprehend them. And he will not find that his soul longs for knowledge of them, inasmuch as he is aware of the impossibility of such knowledge and of there being no gate through which one might enter in order to attain it."

Hundreds of years later, a slow but steady shift in thinking occurred, the many strands of which we now bundle under the term Scientific Revolution. With the advent of modern science, scientists pioneered new fields such as astrophysics and chemistry, and some of the limits to our understanding were thought to be overcome. Of course, this shift was by no means abrupt, with luminaries such as Newton spending much of their efforts on the firmly medieval pursuits of alchemy and mysticism. Nonetheless, there was a distinct move toward the idea that, in principle, the human mind could understand all that it wishes.

This confidence has permeated our approach to scientific insight and in contemporary times has crescendoed in a sort of triumphalism: if we try hard enough, over time we can understand everything. And indeed we have come a very long way in querying the cosmos and finding answers to our questions.

In science, though, we have once again begun to bump up against our limits. It now seems that we would have to do the impossible and travel faster than the speed of light to answer certain questions about the nature of the universe, and we are approaching the limits of understanding certain aspects of our quantum reality. In other words, we are finding questions that we cannot answer. In the words of the biologist J. B. S. Haldane, "Now, my own suspicion is that the universe is not only queerer than we suppose, but queerer than we *can* suppose." While we shouldn't give up on these questions, in place of answers we may end up finding more and more limitations to what we can know.

The same is true of our own creations. We are seeing the limits of our understanding of computing, transportation, medical devices, and so many other technologies we ourselves have made. So as we continue to construct incomprehensible innovations, it might be time to return to the way people thought in an earlier time—a time when it was taken for granted that there is knowledge no human can possibly attain.

What form should this response take? As mentioned in the introduction, as our technologies become more complicated, and we lose the ability to understand them, our responses tend toward two extremes: fear and awe.

Contemplating a fantastically intricate technological system, some of us are overwhelmed by its power and complexity, and respond with fear of the unknown. Others tend toward an almost religious reverence when faced with technology's beauty and power. The video game designer and writer Ian Bogost has even suggested that replacing the term "algorithm" with the word "God" changes little of what is being said about technology in today's discourse.

But technology, while it suffuses our society, is not the product of a perfect and immaculate process. Technologies are kluges. They are messes cobbled together over time from many pieces, and while they are indubitably exciting, they do not merit unquestioning wonder or profound existential concern.

Neither fear nor awe is a productive response; both cut off questioning and the potential for gaining even a hint of understanding. While some caution is necessary when dealing with anything that is more than moderately complicated, fear in the face of what we don't understand abdicates the responsibility to delve deeper and understand what we can, even just a bit.

Unfortunately, the other extreme—the worship of technology—creates the same problem. If we view an algorithm or technology as far more beautiful and impressive than it actually is, this reverence also cuts off further questioning.

In between lies a third path: humility. We must have humility without reverence, and curiosity without fear. The computer scientist Edsger Dijkstra has even written of the "humble programmer" who respects "the intrinsic limitations of the human mind." Even if we could eliminate our mental biases, or massively increase our brainpower, we are still ultimately finite beings. And in the face of this finitude, humility is different from both hubris and humiliation. Humility recognizes our own limitations but is not paralyzed by them, nor does it enshrine them. A humble approach to our technologies helps us strive to understand these human-made, messy constructions, yet still yield to our limits. And this humble approach to technology fits quite nicely with biological thinking. While at every moment an incremental approach to knowledge provides additional understanding of a system, this iterative process will always feel incomplete. And that's okay.

New York Times columnist David Brooks has noted, "Wisdom starts with epistemological modesty." Humility, alongside an interest in the details of complex systems, can do what both fear and worship cannot: help us peer and poke around the backs of our systems, even if we never look them in the face with complete understanding. In many instances, an incomplete muddle of understanding may be the best that we can do. But it's far better than nothing.

Both reverence and fear tempt us to throw our hands in the air and give up. That we cannot afford to do. We must continue to strive to understand these systems. Humility simply means accepting that scientific triumphalism is misplaced: we can never achieve complete or perfect understanding. And if we accept that, perhaps no longer will our soul long for such knowledge, as Maimonides said. Then we can become more philosophical and less dismayed about our failures to understand. When even software experts recognize that some computer bugs are simply in the realm of "metaphysics," it is time for all of us to reconcile ourselves with humility in the face of technology.

In fact, humility and muddling through are noble choices when confronting our complex technologies. As role models we might take the subset of scientists who look at an organism's genome, rife with sections that may not have any purpose whatsoever—having evolved through a bizarre accretive process much as technology did—and nevertheless see a "glorious mess." I find the juxtaposition of "glorious" and "mess" profoundly biological, but also filled with humble admiration. Our technologies are messes, and we can never divine their entirety, but recognizing that they are "glorious messes" is a powerfully optimistic feeling. Details and imperfect understanding can overwhelm us, or make us giddy with excitement. They may never lead to a profound understanding of the entire system, but that's fine.

John Gall, a retired pediatrician, is the author of a book called *The Systems Bible.* Originally titled *General Systemantics,* first published in 1975 and now in a revised and expanded third edition, it is a playful exploration of how to approach complex systems—though Gall uses the term more broadly, encompassing social systems as well as those technological systems we have constructed. The book includes maxims such as "The ghost of the old system continues to haunt the new," "The system always kicks back," and the Unawareness Theorem: "If you're not aware that you have a problem, how can you call for help?" Gall's rules and analyses of systems are insightful and fun. And, as you might expect from its original title, the book's thesis is that systems are prone to antics—they do things we don't expect, they bite back—and it's quite hard to eliminate that behavior.

As some of these maxims might suggest, Gall makes a number of points similar to those made here—systems accrete, expand "beyond human capacity to evaluate," and are subject to unexpected behavior—but I am interested in a particular few. Gall notes that it is easier to work with what you have than to redesign something from scratch: the latter will likely cause more problems than you expect. And if you do feel that you have to create a whole new system, make it a small one, if possible. There are ways to mitigate the failures of systems, at least somewhat.

Ultimately, Gall's maxims and aphorisms seem to me to boil down to one perspective: humility in the face of systems that are so difficult to design, redesign, or rebuild from the ground up. These systems, no matter what their origin or function, will ultimately take similar unwieldy forms, and in our efforts to understand and control them, we must be comfortable with muddling through. When we fully grasp that our systems will always become complex, we will be better prepared to build them from the outset, and better able to recognize and even revel in the surprises and complications when they kick back, as they surely will.

A humble approach to complex technology will serve us well. And one of the key features of this intellectual humility is an insight gleaned from biological thinking: that glimpses into the massive inner workings of these complex systems—little gateways into the machine—may be the best we can do, and that they can be enough.

Glimpses Under the Hood

When the designer Don Norman was backing up his computer to a server, he sat back and watched its progress, reading what it was doing at each step. At one point, Norman noticed that the computer program had reached the stage where it was "reticulating splines." This phrase sounded complicated, and that was reassuring to Norman—this program must really know what it was doing. But he didn't. He got curious, and after some research he discovered—as any good fan of SimCity 2000 would know—that this was actually an inside joke, a nonsensical phrase inserted into the game that only sounds like it means something. Ever since, it has cropped up in various games and other software.

Think back to the last time you installed a new piece of software. Did you know what was going on? Did you clearly understand where various packages were being placed in the vast hierarchy of folders on your hard drive, and what bits of information were being modified based on the specific nature of your computer and its operating system?

Unlikely. Rather, you monitored the progress of this installation by watching an empty rectangle slowly fill over time: a *progress bar.* This small interface innovation was developed by the computer scientist Brad A. Myers, who initially called these bars "percent-done progress indicators" when he created them as a graduate student. They seem to soothe users by providing a small window into an opaque process. Is a progress bar completely accurate? Probably not. Sometimes progress bars are almost completely divorced from the underlying process. But for the most part, a progress bar and other design decisions—such as a bit of text that describes what is happening during a software installation—can provide a reassuring glimpse into a vast and complicated process.

More and more, we have constructed user interfaces that abstract away complexity, or at least partially shield it from the user, bringing together the fields of complexity science and user interface design. Whether in our computers, our cars, or our appliances, these technologies lower a veil between us and how they operate. And rather than grapple with the increasingly byzantine tax code, many of us use the friendly user interface of TurboTax. Yet behind this software is an enormously complicated set of laws and regulations, rendered into the computer code of if-statements and exceptions.

But as long as we have small ways of maintaining some intuition of what is going on beneath the surface—even if it's not completely accurate—we can help users avoid an unnerving discomfort with the unknown.

My family's first computer was the Commodore VIC-20, billed by its pitchman, *Star Trek*'s William Shatner, as "the wonder computer of the 1980s." I have many fond memories of this antiquated machine. I used to play games on it with cassette tapes that served as primitive storage devices. One of the cassettes we owned was a *Pac-Man* clone that my brother and I played a lot. Instead of a yellow pie with a mouth, it featured racing cars.

But we also had games whose code we typed in ourselves.

While you could buy software for the VIC-20 (like the race-car game), a major way that people acquired software in those days was through computer code published in the pages of magazines. Want to play a fun skiing game? Then type out the computer program into your computer, line by line, and fire it up for yourself. No purchase necessary. These programs were common then, but no longer. The tens of millions of lines of code that make up today's game software would fill far more than one magazine.

Typing code into our computer brought us closer to the machine. I saw how bugs occurred—I have a memory of that skiing program creating graphical gibberish on one side of the screen, until the text was corrected—and I also saw that there was a logic and texture to computer programs. Today's computer programs are mysterious creations delivered whole to the user, but the old ones had a legible structure.

Later in the 1980s, my family abandoned Commodore for Apple, and I have used some kind of Macintosh ever since. Our first Mac was something incredible to childhood me. I was entranced by the mouse, and the games, such as *Cosmic Osmo,* which offered rich, immersive realms that you could explore just by clicking. These early Macintoshes could even speak, converting text to speech in an inhuman monotone that delighted my family. The presentation Steve Jobs made introducing the Macintosh in 1984 is profoundly emotional and impressive to watch. And yet, something was lost in my family's rush to embrace the Mac's wonders. We became more distant from the machine. We see this trend continuing even today with the iPad, so slick and pristine that I don't even know how files in it are stored.

However, I had HyperCard for our Mac. HyperCard was this strange combination of programming language and exploratory environment. You could create virtual cards, stitch them together, and add buttons and icons that had specific functionality. You could make fun animations and cool sounds and even connect to other cards. If you've ever played the classic game *Myst,* it was originally developed using HyperCard. HyperCard was like a prototypical series of web pages that all lived on your own computer, but it could also do pretty much anything else you wanted. For a kid who was beginning to explore computers, this visual authoring space was the perfect gateway to the machine.

One program I built with HyperCard was a rudimentary password generator: it could make a random string you could use as a password, but it also had options to make the random passwords more pronounceable, and hence more memorable over the long term. It was simple, but definitely ahead of its time, in my unstudied opinion.

The computer game designer Chaim Gingold calls gateways like HyperCard "magic crayons." Like the crayon in the children's book *Harold and the Purple Crayon* that allows the young hero to draw objects that immediately take on reality, magic crayons are tools that, in Gingold's words, "allow nonprogrammers to engage the procedural qualities of the digital medium and build dynamic things." Even in the Apple world, commonly viewed as sterilized of messy code and computational innards, HyperCard allowed access to the complex powerhouse of the digital domain. HyperCard provided me with the comfort to enter this world, giving me a hint of the possibilities of working under the hood.

All complex systems that we interact with have different levels that we can examine, created in technology by the deliberate abstractions we construct and in nature by the abstracting powers of scale and evolution. In biology, we can zoom up from biochemical enzymes to mitochondria to cells to organs to whole creatures, even entire ecosystems, with each level providing different layers of insight. As we abstract up from one level to the next, we lose fine-grained control and understanding, but we are also able to better comprehend the larger-level system. In computer software, we can move up from individual bits and bytes to assembly language to higher level computer code to the everyday user interface that allows us to click on, drag, and use a web browser. Each successive level brings us more functionality, but it also takes us further away from the underlying logic of the machine.

Of course, as should be clear by now, it's unlikely that that logic will ever be entirely comprehensible. But we should be able to glimpse under the hood a little. If we see our tablets and phones as mere polished slabs of glass and metal, performing veritable feats of magic, and have little clue what is happening beneath the surface or in their digital sinews, something is lost. In fact, this can cause problems: when our systems are so completely automated, we have little ability to respond when something goes wrong. This problem of being shielded from the inner workings of the technology around us has been called "concealed electronic complexity": mind-boggling complexity lies within our devices but is entirely hidden from our view.

In the 1960s, a component of the telephone system was designed in such a way that when it detected that it had failed, it simply connected the user to a wrong number. This redirected people to blame human error—users would think they had simply misdialed—rather than confront the fallibility of the technology itself. Without the knowledge of what was really happening, the person who dialed had a different sense of the system's authority and mystery than she would have if she had seen it clearly as the complicated yet imperfect construction it was.

We need glimpses under the hood to see, even if incompletely, what is going on. When these technologies exceed our ability to fully understand them, such glimpses will matter to the expert as well as to the average user of a technology. It's not enough just for one person, or even a handful of individuals, to see under the hood and recognize the limits of our systems and ourselves. We can't cede this responsibility. Each of us needs to pay attention to these glimpses. Without them, we drift away from a humble but vigilant intuition about these systems and toward reverence or fear. Being able to peek underneath the hood of technology isn't just interesting or educational; it helps inoculate us against unhealthy perspectives toward our technologies. In the Entanglement, we need this protection more and more.

But what if we can't easily get glimpses under the hood? What if a system is so incredibly sophisticated that these little windows either are too difficult to construct or provide too little insight? There is another approach. Simulations are a way to provide us with the beginnings of intuition into how a complex technology works.

While we can't actually control the weather or understand it in all its nonlinear details, we can predict it reasonably well, adapt to it, and even prepare for it. Weather models are incredibly complicated, though each individual part is still designed to be understandable. We look to these models to plan our wardrobe and our activities of the day and week, but also to get a sense, even if an imperfect one, of how the atmosphere operates. And, of course, when the outdoors delivers us an unexpected blizzard or deluge, we manage as best we can.

Just as we have weather models, we can begin to make models of our technological systems, even somewhat simplified ones. Playing with a simulation of the system we're interested in—testing its limits and fiddling with its parameters, without understanding it completely—can be a powerful path to insight, and is a skill that needs cultivation.

For example, the computer game SimCity, a model of sorts, gives its users insights into how a city works. Before SimCity, I doubt many outside the realm of urban planning and civil engineering had a clear mental model of how cities worked, and we weren't able to twiddle the knobs of urban life to produce counterfactual outcomes. We probably still can't do that at the level of complexity of an actual city, but those who play these types of games do have a better understanding of the general effects of their actions. We need to get better at "playing" simulations of the technological world more generally, teaching students how to play with some system, examining its limits and how it works, at least "sort of." This play—tweaking a simulation of technological failure and seeing how it responds—can provide a greater comfort with large and unwieldy systems and can help us as we move forward through this world of increasingly complicated technology.

We also need interpreters of what's going on in these systems, a bit like TV meteorologists. Near the end of *Average Is Over,* the economist Tyler Cowen speculates about this new breed of future interpreters. He says they "will hone their skills of seeking out, absorbing, and evaluating this information. . . . They will be translators of the truths coming out of our networks of machines. . . . At least for a while, they will be the only people left who will have a clear notion of what is going on." These interpreters—who will likely be comfortable with simulations—can help provide us with a glimmer of intuition into complex systems. As more and more of our technology becomes impenetrable, schools and universities will need to incorporate playing with simulations and creating simplifying intuitions as necessary skills—skills in the domain of the science of complexity—equipping each of us to become our own interpreter of these systems.

Simplifying intuitions—the kind of general models developed by complexity scientists—can lay the groundwork for heuristics on how to approach these systems without getting overwhelmed by their complexity. Along with muddling through—the perspective of John Gall's "systemantics"—these little windows into massively complex systems embody the humble approach to technology that we need to adopt.

And there is yet one more positive mind-set we can resort to when we face complex technological systems.

Fostering *Naches*

The Yiddish language possesses a wonderful word, *naches*. It means pride or joy. To *shep naches,* as it's said in Yiddish, is generally to derive vicarious pride from the accomplishments of those close to you, especially your children. It is one of the purest pleasures, and one that you hear spoken of during bar mitzvahs, weddings, and graduations.

Immigrants feel it is important for the next generation to be better off, and for the generation after that to positively thrive. For parents, their offspring must always be more intelligent and more successful than they are, and the drive to make this happen is no doubt part of the reason why we have such *naches.* So, why should we not *shep naches* from the accomplishments of our technologies?

We might not understand what these machines or systems are doing. In some cases, it was even other pieces of technology that constructed them. But they are our brainchildren, our intellectual offspring, and they, too, can give us *naches.* Humans have valued this emotion for thousands of years, and it brings us great happiness. We just need to transfer our parental pride to the technological realm.

So what does *naches* mean exactly for technology? At the most basic level, the creators of these machines can *shep naches* from the accomplishments and powers of their technological progeny. Among all their other achievements, computer programs are now even capable of generating sophisticated artworks or musical compositions. The composer David Cope of the University of California, Santa Cruz, has developed software that generates novel musical compositions in the style of a given composer. And they sound really good. A Scott Joplin–style composition sounds like Joplin. While Cope didn't create these songs directly, he still can take pride in their construction. His computational creations can provide him with *naches.* Similarly, the creators of IBM's Watson might *shep naches* from the machine's win over its human opponents on *Jeopardy!*

We can broaden this sense of *naches* still more. Many of us support a sports team and take pride in its wins, even though we had nothing to do with them. Or we become excited when a citizen of our country takes the gold in the Olympics, or makes a new discovery and is awarded a prestigious prize. So, too, should it be with our machines for all humanity: we can root for what humans have created, even if it wasn't our own personal achievement and even if we can't fully understand it. Many of us are grateful for technological advances, from the iPhone to the Internet, even if we don't know how they work.

If our creations' complexity outstrips our ability to understand them, we should not be disturbed or disappointed. When our children do something surprising and amazing that we can't really understand, we don't despair or worry; we are delighted and even grateful for their success. We can respond similarly to our technological creations.

But *naches* is also a framework for helping us recognize that we are following the same trajectory we have been tracing for thousands of years, in which fewer and fewer people are able to understand the most complex parts of the world we have created. These recent trends are not really so new or different from what has come before.

Humility refracted through the lens of *naches* is an optimistic perspective on the incomprehensibility to which we have given birth. Furthermore, this humility is a recognition of the balance between our unceasing desires—and in the case of experts, almost a mandate—to grasp the dimensions of what we have constructed and our limits in doing so.

Humility can also clarify a distinction between mystery and wonder.

Mystery and Wonder

A book published in the second half of the nineteenth century called *The World of Wonders* is a grab bag of intriguing ideas, fascinating things, and strange events. It is written in the manner of someone delighted with life and constantly curious, interspersing anthropic advances with phenomena from nature. It includes "all that is most wonderful in history and philosophy and the marvels of science, the wonders of animal life revealed by the glass of the optician or the labours of the chemist," and much more. It is a marvel of the Victorian period and demonstrates the many ways that we can find wonder all around us. But this wasn't wonder at a world that was unexplained. It was wonder precisely because our advances and technologies were so interesting, so powerful, and above all, understandable—not mysterious.

There are two potential extreme responses to mystery in our complex technological systems. The first is to underemphasize it, to such a degree that there are no mysteries. We tell ourselves simplifying stories about how things work, stories that, while appealing, severely underestimate the complexity of our systems. We say that we understand a technology we are using, and any issue is just a small glitch that needs to be ironed out. Many large technology companies tend toward this style, dismissing any unexpected behaviors as minor issues in the process of being eliminated.

At the other extreme are those who actively seek out mystery and the unknown, whether or not it's actually there. These people—often laypeople—wear their mystification at a device's or system's inner workings almost as a badge of honor, declaring their iPhones or the ways of the electrical grid to be nearly magical. These are the kindred spirits of Calvin, friend to the philosophical tiger, who asked his father how lightbulbs and vacuum cleaners worked, and was content with the answer: "Magic." Many of us, though, lie somewhere in the middle. We might recognize a certain amount of mystery, yet still want the world to be explainable.

Things get messier when we bring in wonder, our ability to marvel and to feel a sense of the numinous in the world around us. We may delight in the complexity or beauty of a phenomenon, but, as is clear from *The World of Wonders,* wonder doesn't have to be at the expense of understanding—or vice versa. In fact, many people find it more gratifying the more they can grapple with and succeed in understanding a really complicated system. Don Norman has written of the delight in truly understanding the infield fly rule in baseball. In a complicated rule set—baseball's rules are over 200 pages long—just the definition of an infield fly contains a thicket of exceptions and qualifiers: "An infield fly is a fair fly ball (not including a line drive nor an attempted bunt) which can be caught by an infielder with ordinary effort, when first and second, or first, second and third bases are occupied, before two are out." Fans enjoy working hard to understand this complex system, and find wonder in it.

So how do we balance mystery and wonder when it comes to these vast systems of our own making, specifically those that are—no matter how hard we try—not fully comprehensible?

We mustn't give in to the temptation to collapse these two senses together—to conflate mystery and wonder. Those who do this argue that the unexplained must necessarily inspire us. The mysterious is incorrectly identified with the wondrous: something not understood should therefore be marveled at. Similarly, we must resist the temptation to abandon wonder once we understand something completely. This happens too often with technological change. Changes are happening so rapidly that we forget to marvel at how impressive our understanding of the universe—and our ability to harness it—has become. We forget how recently we gained the ability to render three-dimensional worlds on our screens, communicate instantly across the planet, or even summon decades-old television commercials with the click of a mouse. The knowledge that has made these changes possible too often fails to inspire wonder. Then we are left with "the sad inertness of a world explained and controlled," as the philosophers Hubert Dreyfus and Sean Dorrance Kelly describe the disenchantment that threatens when wonder declines as our technological powers rise—driving some into fantasies of an enchanted past.

Instead, we must work to maintain two opposing states: mystery without wonder and wonder without mystery. The first requires that we strive to eliminate our ignorance, rather than simply reveling in it. And the second means that once we understand something, we do not take it for granted.

In essence, these two final combinations are what I think of as the scientific mind-set, the prerequisite for our ability to learn new things and solve the puzzles that confront us. When we wonder too much at the mysteries we have inadvertently created, we undercut our desire to eliminate those mysteries and understand our Entanglement.

We will never fully understand it. The kluges are too

complex. We will always be left with some mystery, but that's okay. As long as we neither fear nor revel in it, we can take the proper perspective: humility, even with a touch of awe. Because we *built* these systems. Humility, wonder, and even some *naches* might be the best that we can do. We must continue to pursue a biological mind-set toward these systems, even if we don't gain a full understanding. We can be humble if we fail, but there are paths for our intuition, glimpses into our technologies.

When vehicles manufactured by Toyota began accelerating unexpectedly, even resulting in deaths, experts could not figure out exactly what had caused the error. One suggestion was to inform Toyota drivers that if their cars ever began to accelerate uncontrollably, a simple solution involved shifting the car into neutral. Of course, when you are a consumer entrusting your life and your family's lives to a powerful machine that might behave in an unpredictable and frightening manner, that's not what you want to hear. It sounds almost like a callous Band-Aid slapped onto a deeper problem.

If we know that rare failures and unexplainable glitches are by-products of the evolution of complexity, however, and that even the most vigilant engineering cannot anticipate them all, we will be better prepared to respond, should they occur. In that case, "Just shift into neutral" isn't the worst thing to tell someone who is about to step into a fantastically complicated machine, if they accept that these messy, incomprehensible systems are the new reality. Both designers and drivers are now partners in pioneering the Entanglement and are not as far apart as they might seem, at least in one important respect: neither can fully predict the behavior of a complex system. Many of us already recognize this, albeit in a much more minor way and even if only subconsciously, smacking a machine in frustration or simply turning our computers off and then on again, in the hope that the complexity will resolve itself in our favor.

The twentieth century brought us numerous *limitative theorems,* statements that placed bounds upon what we could ever know and understand. Kurt Gödel showed that in mathematics there will always be statements that can never be proven as either true or false, within a given mathematical system. In computer science, Alan Turing demonstrated the limits on what any machine could ever do, no matter how fancy an algorithm one might develop. But neither of these fields died out or suffered a huge setback. Despite being bounded by limits, they flourished, in many ways far beyond what these men could have ever imagined.

In building and using complex technological systems, there are limits to what we can understand about how they work and how they fail. This does not mean an end to such creations.

Rather, it simply means that as we continue to build such systems, ones that will continue to grow ever more impressive and sophisticated, we must recalibrate our expectations.

Starting from the recognition that we can't fully understand these systems changes how we approach them. If you're involved in their construction, you can assemble them with an awareness of our limits of comprehension. If you interact with such a system, you can recognize that bugginess and unexpectedness are common, not the exceptions, and that if you try to eliminate them, not only will you fail, but you may even make things worse.

The science fiction writer William Gibson has spoken of the "unthinkable present," the setting for many of his stories. This phrase contains the kernel of what we are currently living inside of, this Entanglement. But in the end, these systems are not completely incomprehensible, nor will their properties and qualities remain forever ineffable. The story doesn't end with horror at the unknowable or wonder at a completely impenetrable mystery. Humility tempered with the acknowledgment of an iterative and biological approach to understanding must guide our interactions with our overcomplicated technology. But no matter the details of our perspective, we do not have to give in to the logic of Ludwig Wittgenstein: "What we cannot speak about we must pass over in silence."

There is much to be said about what we can't fully understand and how to respond to it. We can live and thrive in this Entanglement.

Further Reading

This book did not emerge whole cloth. Its ideas grew out of and were drawn from a very large number of sources. In addition to what I reference in the endnotes, I'd like to highlight several books and articles that were particularly thought-provoking and helped inform some of my thinking when writing this book. Some of these references aren't even discussed in this text but could be useful if you are interested in delving further into the topics I touch upon.

In addition, there are many topics that, while relevant, I did not discuss in detail in this book. These include such things as technological change and how that affects education, more on Big Data, our increasing partnership with our machines, automation and the future of jobs, and the many, many different types of systems that we see around us, including those in manufacturing, food, government, energy, and so much more. Some of the works below address these topics.

The Systems Bible by John Gall is a bizarre romp into how to think about large systems and how they work, or don't. It is fascinating and much of my thinking parallels Gall's.

Out of Control by Kevin Kelly includes some of the same points about biological thinking and how technology is becoming increasingly biological and unable to be understood, though from the perspective of emergence and biological complexity and the use of biological principles to build technologies. I also recommend *What Technology Wants* by the same author.

Autonomous Technology by Langdon Winner. The penultimate chapter is particularly salient and raises many of the same issues discussed in this book.

The Ingenuity Gap by Thomas Homer-Dixon is about how to deal with an increasingly complex and unpredictable world. It explores all manner of systems, many of those beyond the scope of this book, and is fascinating.

Normal Accidents by Charles Perrow is a classic text on failure and "living with high-risk technologies."

Geek Sublime by Vikram Chandra is a beautiful meditation on the nature of programming and computers, among many fascinating topics.

Infinite in All Directions by Freeman Dyson includes some

great discussions on the nature of scientific thinking and discovery.

The Mythical Man-Month by Frederick Brooks is a classic exploration of software development and design. It focuses on the management of engineering teams but has huge amounts of interesting thoughts on the nature of software and programming more generally.

Curiosity: How Science Became Interested in Everything by Philip Ball is a great account of how scientists, especially the early ones, began exploring our incredibly diverse and detailed world and attempted to impose upon it a sense of order.

Living with Complexity by Don Norman examines the origins of (and need for) complexity, particularly from the perspective of design.

The Techno-Human Condition by Braden R. Allenby and Daniel Sarewitz is a discussion of how to grapple with coming technological change and is particularly intriguing when it discusses "wicked complexity."

Superintelligence by Nick Bostrom explores the many issues and implications related to the development of superintelligent machines.

The Works, The Heights, and *The Way to Go* by Kate Ascher examine how cities, skyscrapers, and our transportation networks, respectively, actually work. Beautifully rendered and fascinating books.

The Second Machine Age by Erik Brynjolfsson and Andrew McAfee examines the rapid technological change we are experiencing and can come to expect, and how it will affect our economy, as well as how to handle this change.

The Glass Cage by Nicholas Carr is about the perils of automation and the related technological complexity around us.

Shop Class as Soulcraft by Matthew B. Crawford explores the importance of getting close to our technologies again, as part of the virtue of manual labor.

Summa Technologiae by Stanisław Lem (translated by Joanna Zylinska) is a wide-ranging exploration from the 1960s by a science fiction writer of the future of technology, with an emphasis on the limits of humanity's powers of understanding.

Think Twice by Michael Mauboussin looks at how to think properly—and often counterintuitively—about the complex systems that are all around us.

"When Technology Ceases to Amaze" by Robert Herritt in *The New Atlantis* 41, Winter 2014, pages 121–31, is a great essay about technological wonder, complexity, and amazement.

"In the Beginning . . . Was the Command Line" by Neal Stephenson, an essay and also a short book, is essentially a long and winding meditation on computing. Though outdated, it contains a great deal of wisdom on connection to and detachment from technology.

"I, Pencil" by Leonard E. Read is a brief essay that explores the highly interconnected socioeconomic system involved in manufacturing a pencil, the totality of which no single individual understands.

I also recommend looking at the writings of Edsger Dijkstra in the E. W. Dijkstra Archive, many of which are classics in computer science and border on philosophical musings on the nature of technology. Available online at https://www.cs.utexas.edu/~EWD/.

For additional reading, please go to arbesman.net to find essays of mine on these topics.

Acknowledgments

First of all, thank you to my editor, Niki Papadopoulos, for the invaluable support throughout the entire book-writing process. Max Brockman, thanks for helping me shape the idea for this book.

Portions of this book have been developed elsewhere, as well as appearing in modified form in numerous places, including *Aeon, Nautilus, Slate, Wired* Opinion, Edge.org's "The Edge Question—2015: What Do You Think About Machines That Think?" and *Arc.* Thank you to these outlets for providing me with an opportunity to work out my ideas in public. I'd also like to thank *Wired* Science for having let me develop many of the ideas in this book on my blog there, where they appeared in a considerably more incoherent form. Also, those of you who subscribe to my email newsletter were able to experience early versions of some of these topics as well.

Thank you to all my early readers and interlocutors for your generosity of time and wealth of expertise, both of which are quite rare in this modern era. These include Josh Arbesman, Zev Berger, Andrew Blum, Aaron Clauset, Lori Emerson, Joshua Fairfield, Henry Farrell, Laurence Gonzales, Edward Jung, Aaron Kahn, Dan Katz, Mykel Kochenderfer, Steven Miller, Megan Owen, Elnatan Reisner, Nahum Shalman, Jacob Sherman, Ted Steinberg, Brian Stephens, Harry Surden, and Jevin West. To all of you, and anyone else who helped shape the ideas found here, many thanks for helping make this book better, and preventing some pretty ridiculous goofs. That being said, all errors remain my own. I'd also like to single out in particular the philosophy research group at the University of Kansas that examines human understanding and software. As a visiting scholar in the philosophy department at KU, I am a part of this small group, and all the feedback and ideas from the members of this group—John Symons, Jack Horner, Ramon Alvarado—have improved my thinking on the topics of this book immeasurably. David Steen generously helped guide my thinking about wildlife ecology and field biology. Josh Sunshine provided a great deal of support and advice, fielding an innumerable number of questions related to computer science and software engineering. Michael Vitevitch was invaluable with his expertise and guidance on the nature of language and how we process it. Michael Barr and Philip Koopman also provided useful advice on the problem of unintended acceleration and technological complexity in Toyotas.

I am also indebted to the Ewing Marion Kauffman Foundation, which provided me with a great environment for writing much of this book. Thank you to the Foundation and my colleagues there for your incredible feedback and warmth, especially in letting me work in not too monk-like a fashion on this text.

In addition, the team at Lux Capital has been enormously supportive of this project, and provided numerous insights and feedback. Thanks so much to you all.

Finally, my family. My parents provided much-needed support and advice, as well as detailed feedback on many drafts, and my grandfather is the best sounding board I know. I am incredibly grateful to you all. To my wife, Debra, you gave me the space and time to write this book and make sure that it was the best version possible. Your patience during this process, from reading early drafts to hearing me rant about technology, has been astonishing and wonderful.

And Abigail: you love books, though currently you are enamored of those that are considerably shorter and often have sturdier pages than this one. Your ability to bring joy to me every day is something that I still find incredible. Your support throughout this book from its inception—even though it consisted primarily of hugs, cuddles, giggles, and delightful chatter—is more than I ever could have hoped for. While I hope this book finds a broad audience, ultimately it's written for you and your new brother, Nathan. You both will come of age in a world that I can only begin to imagine, one far different from the one we live in today and far less understandable. This book hopes to explain, in a small way, why you shouldn't be frightened by this future.

Notes

INTRODUCTION

1 **a lot of buggy software:** Some speculate that the WSJ.com outage was caused by an overload when subscribers thronged to read about the NYSE shutdown. Jose Pagliery, "Tech Fail! Explaining Today's 3 Big Computer Errors," *CNN* Money, July 8, 2015, http://money.cnn.com/2015/07/08/technology/united-nyse-wsj-down/. For more on this glitch and what it portends, see Zeynep Tufekci, "Why the Great Glitch of July 8th Should Scare You," *The Message,* July 8, 2015, https://medium.com/message/why-the-great-glitch-of-july-8th-should-scare-you-b791002fff03#.cd6hchnur.

1 **As one security expert stated:** Andrea Chang and Tracey Lien, "Outages at NYSE, United Airlines, WSJ.com Expose

Digital Vulnerabilities," July 8, 2015, http://www.latimes.com/business/technology/la-fi-tn-technical-problems-united-nyse-20150708-story.html.

2 **the particle accelerator in Strasbourg:** Thomas Homer-Dixon, *The Ingenuity Gap: Facing the Economic, Environmental, and Other Challenges of an Increasingly Complex and Unpredictable Future* (New York: Alfred A. Knopf, 2000; repr. Vintage, 2002), 171.

3 ***radical novelty:*** Edsger W. Dijkstra, "On the Cruelty of Really Teaching Computing Science," E. W. Dijkstra Archive: The manuscripts of Edsger W. Dijkstra, 1930–2002, document no. EWD1036, December 1988, http://www.cs.utexas.edu/users/EWD/ewd10xx/EWD1036.PDF [hand-printed original], http://www.cs.utexas.edu/users/EWD/transcriptions/EWD10xx/EWD1036.html [typed transcript]. Dijkstra made a similar point in his Association for Computing Machinery (ACM) Turing Lecture in 1972. E. W. Dijkstra, "The Humble Programmer," *Communications of the ACM* 15, no. 10 (1972): 859–86. Not every technology is computational, but Dijkstra's insight does impinge on much of our technological life.

3 **the *Anthropocene*, the Epoch of Humanity:** For further reading, see Lee Billings, "Brave New Epoch," *Nautilus* 009: January 30, 2014.

4 **a journal article in *Scientific Reports:*** Neil Johnson et al., "Abrupt Rise of New Machine Ecology Beyond Human Response Time," *Scientific Reports* 3:2627, September 11, 2013.

4 **with humans on the sidelines:** "Back in 2008, when it first occurred to Brad Katsuyama that the stock market had become a black box whose inner workings eluded ordinary human understanding . . ." Michael Lewis, "The Wolf Hunters of Wall Street," *The New York Times Magazine,* March 31, 2014, http://www.nytimes.com/2014/04/06/magazine/flash-boys-michael-lewis.html.

5 **This phenomenon of "algorithm aversion":** Berkeley J. Dietvorst et al., "Algorithm Aversion: People Erroneously Avoid Algorithms After Seeing Them Err," *Journal of Experimental Psychology: General* 144, no. 1 (2015), 114–26.

6 **When we delight at Google's brain:** As an example of this kind of reverence, a journalist wrote of feeling "almost levitated" when seeing a Google data center. Stephen Levy, "Google Throws Open Doors to Its Top-Secret Data Center," *Wired,* October 17, 2012, http://www.wired.com/2012/10/ff-inside-google-data-center/. Another writer described the awe that can be induced by Facebook's news feed algorithm: "The news feed algorithm's outsize influence has given rise to a strand of criticism that treats it as if it possessed a mind of its own—as if it were some runic form of intelligence, loosed on the world to pursue ends beyond the ken of human understanding." Will Oremus, "Who Controls Your Facebook Feed," *Slate,* January 3, 2016, http://www.slate.stfi.re/articles/technology/cover_story/2016/01/how_facebook_s_news_feed_algorithm_works.html.

6 **because of its creation by some perfect, infinite mind:** "The worship of the algorithm" is discussed further in Ian Bogost, "The Cathedral of Computation," *The Atlantic,* January 15, 2015, http://www.theatlantic.com/technology/archive/2015/01/the-cathedral-of-computation/384300/.

CHAPTER 1: WELCOME TO THE ENTANGLEMENT

9 **the *Challenger* disaster:** James Gleick, "Richard Feynman Dead at 69; Leading Theoretical Physicist," *The New York Times,* February 17, 1988, http://www.nytimes.com/books/97/09/21/reviews/feynman-obit.html.

10 **car began accelerating uncontrollably:** For further information on "unintended acceleration" in Toyota vehicles, see Ken Bensinger and Jerry Hirsch, "Jury Hits Toyota with $3-million Verdict in Sudden Acceleration Death Case," *Los Angeles Times,* October 24, 2013, http://articles.latimes.com/2013/oct/24/autos/la-fi-hy-toyota-sudden-acceleration-verdict-20131024; Ralph Vartabedian and Ken Bensinger, "Runaway Toyota Cases Ignored," *Los Angeles Times,* November 8, 2009, http://www.latimes.com/local/la-fi-toyota-recall8-2009nov08-story.html#page=1; Margaret Cronin Fisk, "Toyota Settles Oklahoma Acceleration Case After Verdict," *Bloomberg Business,* October 25, 2013, http://www.bloomberg.com/news/articles/2013-10-25/toyota-settles-oklahoma-acceleration-case-after-jury-verdict; Associated Press, "Jury Finds Toyota Liable in Fatal Wreck in Oklahoma," *New York Times,* October

25, 2013, http://www.nytimes.com/2013/10/25/business/jury-finds-toyota-liable-in-fatal-wreck-in-oklahoma.html.

10 **computer scientist Philip Koopman:** Philip Koopman posted a talk with slides, "A Case Study of Toyota Unintended Acceleration and Software Safety," on his blog *Better Embedded System SW,* October 3, 2014, http://betterembsw.blogspot.com/2014/09/a-case-study-of-toyota-unintended.html. See also the report by Michael Barr of the Barr Group on *Bookout v. Toyota,* http://www.safetyresearch.net/Library/BarrSlides_FINAL_SCRUBBED.pdf.

11 **in this case, unnecessarily complex:** There is a distinction between *inherent complexity* and *accidental complexity.* The former is complexity that is required for a system to operate (including provision for various exceptions and special situations). The latter is closer to overcomplication, the complexity that often arises when a system grows by accretion and tinkering rather than a careful plan.

11 **an Ariane 5 rocket exploded:** The Ariane rocket story is told in Homer-Dixon, *The Ingenuity Gap.* Homer-Dixon based part of his narrative on James Gleick, "A Bug and a Crash: Sometimes a Bug Is More Than a Nuisance," 1996, http://www.around.com/ariane.html (which originally appeared in *The New York Times Magazine,* December 1996). For a similar discussion of proximate causes versus the underlying reasons for such sudden system failures, see Chris Clearfield and James Owen Weatherall, "Why the Flash Crash Really Matters,"

Nautilus 023, April 23, 2015, http://nautil.us/issue/23/dominoes/why-the-flash-crash-really-matters.

12 **Three Mile Island nuclear disaster:** Clearfield and Weatherall, "Why the Flash Crash Really Matters."

12 **the system's massive complexity:** Essentially, the failure in each of these cases was due to endogenous complexity—the complexity that evolves within a large system—rather than just to any specific exogenous shock.

12 **popular narrative of the *Challenger*:** It must be recognized that the *Challenger* accident was more complicated than the streamlined story we are often told about its cause. For example, engineers involved were aware of "the risk of catastrophic failure" of the space shuttle—though, as the following source notes, they could not pinpoint a specific reason—and objected to its launch. At the time, it seems, the engineers knew that "temperature might be a causal factor," but were not certain of it. Wade Robison et al., "Representation and Misrepresentation: Tufte and the Morton Thiokol Engineers on the *Challenger*," *Science and Engineering Ethics* 8, no. 1 (2002): 59–81, https://people.rit.edu/wlrgsh/FINRobison.pdf. Further details can also be found in this oral history of the accident, which indicates that engineers seem to have known the cause of the accident and that this information was given to Feynman to highlight in the hearing: Margaret Lazarus Dean, "An Oral History of the Space Shuttle Challenger Disaster," *Popular Mechanics,* February, 2016, http://www.popularmechanics.com/space/a18616/an-oral-history-of-the-space-shuttle-challenger-disaster/.

12 **Whiggish view of progress:** Philip Ball, "Science Fictions," *Aeon,* October 29, 2012, http://aeon.co/magazine/science/philip-ball-history-science/.

12 **as the historian Ian Beacock writes:** Ian Beacock, "Humanist among Machines," *Aeon,* June 25, 2015, http://aeon.co/magazine/society/why-we-need-arnold-toynbees-muscular-humanism/.

13 **described by the sociologist Max Weber:** Max Weber, "Science as a Vocation," in *From Max Weber: Essays in Sociology,* trans. and ed. H. H. Gerth and C. Wright Mills, 129–56 (New York: Oxford University Press, 1946, repr. 1958), available online: http://anthropos-lab.net/wp/wp-content/uploads/2011/12/Weber-Science-as-a-Vocation.pdf.

13 **systems that can't be grasped in their totality:** This new way that our own creations confound us is echoed in a possibly apocryphal quote from Paul Valéry: "So the whole question comes down to this: can the human mind master what the human mind has made?" Quoted in Langdon Winner, *Autonomous Technology: Technics-out-of-Control as a Theme in Political Thought* (Cambridge, MA: The MIT Press, 1977), 13.

13 **"complicated" and "complex" systems:** This is but one of likely very many distinctions between these two terms.

14 **Imagine water buoys:** Thanks to Aaron Clauset for providing the example of tied-together buoys during a discussion.

16 **the infrastructure of our cities:** In Kevin Kelly's view, "Cities are technological artifacts, the largest technology we make." *What Technology Wants* (New York: Viking, 2010), 81.

16 **could fill encyclopedias:** David McCandless, "Codebases: Millions of Lines of Code," infographic, v. 0.9, *Information Is Beautiful,* September 24, 2015, http://www.informationisbeautiful.net/visualizations/million-lines-of-code/. Assuming an encyclopedia has about 30,000 pages and each page could fit 1,000 lines of code, that means that by some estimates, one version of the Macintosh operating system could fill multiple encyclopedias.

16 **300,000 intersections with traffic signals:** The number of "signalized intersections" in the United States is an estimate from the U.S. Department of Transportation's Federal Highway Administration, http://mutcd.fhwa.dot.gov/knowledge/faqs/faq_part4.htm (last modified October 20, 2015).

16 **Autocorrect, which we often deride:** Gideon Lewis-Kraus, "The Fasinatng . . . Frustrating . . . Fascinating History of Autocorrect," *Wired,* July 22, 2014, http://www.wired.com/2014/07/history-of-autocorrect/.

16 **pages in the federal tax code:** Laura Saunders, "Paper Trail," sidebar to article "Don't Make These Tax Mistakes: Fifteen Common Tax-Filing Errors That Can Cost You Dearly," *The Wall Street Journal,* January 31, 2014.

17 **It exists across a rich spectrum:** For a way to think about different levels of understanding of complex systems, see Stephen Jay Kline, *Conceptual Foundations for Multidisciplinary Thinking* (Stanford, CA: Stanford University Press, 1995), 268.

21 **Lee Felsenstein has told the story:** Erik Sandberg-Diment, "A Computer Comes in from the Cold," *The New York Times,* April

21, 1987, http://www.nytimes.com/1987/04/21/science/personal-computers-a-computer-comes-in-from-the-cold.html.

21 **computer scientist Gerard Holzmann:** Gerard J. Holzmann, "Code Inflation," *IEEE Software* (March/April 2015): 10–13.

22 **"too dense to be knowable":** Philip K. Howard, "Fixing Broken Government: Put Humans in Charge," *The Atlantic,* September 22, 2014, http://www.theatlantic.com/politics/archive/2014/09/fixing-broken-government-put-humans-in-charge/380309/?single_page=true.

22 **the writer Quinn Norton has noted:** Quinn Norton, "Everything is Broken," *The Message,* May 20, 2014, https://medium.com/message/81e5f33a24e1.

22 **Langdon Winner notes in his book:** Winner, *Autonomous Technology,* 290–91.

23 **computer scientist Danny Hillis argues:** Danny Hillis, "The Age of Digital Entanglement," *Scientific American,* September 2010, 93.

25 **Take the so-called Flash Crash:** Nick Bostrom, *Superintelligence: Paths, Dangers, Strategies* (Oxford, UK: Oxford University Press, 2014), 17. It is still not entirely clear, however, what caused the Flash Crash.

27 **Understanding something in a "good enough" way:** See also César Hidalgo, *Why Information Grows: The Evolution of Order, from Atoms to Economies* (New York: Basic Books, 2015).

CHAPTER 2: THE ORIGINS OF THE KLUGE

31 **the Internet first began to be developed:** For more, see Barry M. Leiner et al., "Brief History of the Internet," Internet Society, October 15, 2012, http://www.internetsociety.org/brief-history-internet.

32 **the source HTML of Google's homepage:** See Randall Munrose, "DNA," *xkcd,* November 18, 2015, https://xkcd.com/1605/.

32 ***Slate* interactives editor Chris Kirk:** Chris Kirk, "Battling My Daemons: My Email Made Me Miserable. So I Decided to Build My Own Email App from Scratch," *Slate,* February 25, 2015, http://www.slate.com/articles/technology/technology/2015/02/email_overload_building_my_own_email_app_to_reach_inbox_zero.html.

33 **laws and regulations are technologies:** The analogy between computer code and other kinds of codes—legal, moral—can be pushed quite far, but treating these as distinct types of systems is probably best.

34 **to establish the postal service:** Constitution of the United States of America, Article I, Section 8: "To Establish Post Offices and Post Roads"; "Title 39—Postal Service," *Code of Federal Regulations* (annual edition), revised July 1, 2003, available online: http://www.gpo.gov/fdsys/pkg/CFR-2003-title39-vol1/content-detail.html.

34 **United States Code is far more complicated:** Michael J. Bommarito II and Daniel M. Katz, "A Mathematical Approach to

the Study of the United States Code," *Physica A: Statistical Mechanics and its Applications* 389, no. 19 (2010): 4195–200, http://arxiv.org/abs/1003.4146.

34 **airplane the Wright brothers built:** "Wright 1903 Flyer," NASA Glenn Research Center, accessed June 17, 2015, http://wright.nasa.gov/airplane/air1903.html.

34 **A Boeing 747-400:** "747 Fun Facts," Boeing Commercial Airplanes, https://web.archive.org/web/20111205231111/http://www.boeing.com/commercial/747family/pf/pf_facts.html.

34 **the numbers of individual parts:** Kelly, *What Technology Wants,* 279.

35 **Windows operating system became:** Kelly, *What Technology Wants,* 279; David McCandless, "Codebases: Millions of Lines of Code," infographic, v. 0.9, *Information Is Beautiful,* September 24, 2015, http://www.informationisbeautiful.net/visualizations/million-lines-of-code/.

35 **software application Photoshop:** McCandless, "Codebases."

35 **the American telephone system:** M. D. Fagen, ed., *A History of Engineering and Science in the Bell System: The Early Years (1875–1925),* Technical Publication Department, Bell Laboratories, 1975.

37 **FAA began to examine its computers:** The story of the FAA's concern about Y2K is from Homer-Dixon, *The Ingenuity Gap.*

37 **representing the FAA technicians:** Matthew L. Wald, "Warning Issued on Air Traffic Computers," *The New York Times,* January 13, 1998, http://www.nytimes.com/1998/01/13/us/

warning-issued-on-air-traffic-computers.html, accessed February 6, 2015. According to Homer-Dixon, one of the retired programmers was hired to help fix the problem.

37 **machines at the Internal Revenue Service:** Anne Broache, "IRS Trudges On with Aging Computers," *CNET News*, April 12, 2007. See also John Bodoh, "Tech Timebomb: The IRS Is Still Living in the 1960s," *Washington Examiner*, December 17, 2014, http://www.washingtonexaminer.com/tech-timebomb-the-irs-is-still-living-in-the-1960s/article/2557483.

38 **final space shuttle mission:** "The Shuttle: NASA's IT Legacy," *Information Age*, July 18, 2011, http://www.information-age.com/technology/applications-and-development/1641693/the-shuttle%3A-nasas-it-legacy.

38 **In *The Mythical Man-Month:*** Frederick P. Brooks Jr., *The Mythical Man-Month: Essays on Software Engineering*, anniversary ed. (Boston, MA: Addison-Wesley, 1995; orig. pub. 1975), 53. It seems that this quote is a Latin proverb, misattributed to Ovid.

39 **a process of accretion:** This term is also used in Homer-Dixon, *The Ingenuity Gap*.

39 **"Typically, outdated legacy systems":** Stewart Brand, *The Clock of the Long Now: Time and Responsibility* (New York: Basic Books, 1999), 85.

40 **only gingerly poke it:** From an essay by Stewart Brand: "Beyond the evanescence of data formats and digital storage media lies a deeper problem. Computer systems of large scale are at the core of driving corporations, public institutions, and indeed whole sectors of the economy. Over time, these gargantuan

systems become dauntingly complex and unknowable, as new features are added, old bugs are worked around with layers of 'patches,' generations of programmers add new programming tools and styles, and portions of the system are repurposed to take on novel functions. With both respect and loathing, computer professionals call these monsters 'legacy systems.' Teasing a new function out of a legacy system is not done by command, but by conducting cautious alchemic experiments that, with luck, converge toward the desired outcome." "Written on the Wind," The Long Now Foundation, February 11, 1998, http://longnow.org/essays/written-wind/.

40 **pages of instructions for the 1040:** Christopher Ingraham, "Charted: The Skyrocketing Complexity of the Federal Tax Code," *Washington Post,* April 15, 2015, http://www.washingtonpost.com/blogs/wonkblog/wp/2015/04/15/charted-the-skyrocketing-complexity-of-the-federal-tax-code/.

40 **Supreme Court has ruled:** In *Cheek v. United States,* 498 U.S 192. See also *United States v. Murdock,* 290 U.S. 389.

41 **pages in the *Code of Federal Regulations:*** Susan E. Dudley and Jerry Brito, *Regulation: A Primer,* 2nd ed. (Arlington, VA, and Washington, DC: Mercatus Center, George Mason University, and The George Washington University Regulatory Studies Center, 2012), 5, http://mercatus.org/publication/regulation-primer.

41 **known as Parkinson's Law:** "Parkinson's Law," *The Economist,* November 19, 1955, accessed February 26, 2015, http://www.economist.com/node/14116121. This relationship does not seem to hold when a country is at war.

41 **those in the software world have enshrined this idea:** See Meir M. Lehman, "Programs, Life Cycles, and Laws of Software Evolution," *Proceedings of the IEEE* 68, no. 9 (1980): 1060–76. Also Lehman et al., "Metrics and Laws of Software Evolution—The Nineties View," in *METRICS '97, Proceedings of the Fourth International Software Metrics Symposium* (Washington, DC: IEEE Computer Society, 1997), 20–32. In addition, see *The Systems Bible* by John Gall, discussed later and in the Further Reading section.

42 **rewrite a piece of software:** This rewriting is related to *refactoring:* recoding a piece of software so that internally it is a lot better and cleaner, even though the external functionality remains unchanged.

42 **trade-offs in time, effort, and money:** Astronaut Alan Shepard famously said in a post-flight briefing, "It's a very sobering feeling to be up in space and realize that one's safety factor was determined by the lowest bidder on a government contract."

42 **our cities have gas pipes:** John Kelly, "Look Out Below: Danger Lurks Underground from Aging Gas Pipes," *USA Today,* September 23, 2014, http://www.usatoday.com/story/news/nation/2014/09/23/gas-pipes-cast-iron-deaths-explosions-investigation/15783697/.

42 **run on 1930s technologies:** Metropolitan Transportation Authority (MTA), "CBTC: Communications-Based Train Control," July 20, 2015, https://www.youtube.com/watch?v=Mjx3S3UjmnA.

42 **systems become more complicated over time:** Such systems even have to start off in a complex state, in order to anticipate edge cases and exceptions, as discussed later in the chapter.

44 **described as a "godsend":** Harry McCracken, "Fifty Years of BASIC, the Programming Language That Made Computers Personal," *TIME,* April 29, 2014, http://time.com/69316/basic/. A more principled way of managing such branches and loops is to use more-explicit versions of them, such as "for loops."

45 **but actually "harmful":** Edsger W. Dijkstra, "Go To Statement Considered Harmful," *Communications of the ACM* 11, no. 3 (1968): 147–48. Note that Dijkstra's original title for this paper was "A Case Against the GO TO Statement."

45 **make sense of and impose order:** Refactoring is one such method. Software developers Brian Foote and Joe Yoder have sought to explain why the "de-facto standard software architecture" is a "casually, even haphazardly structured system"—a "Big Ball of Mud"—and to map out ways to improve such systems from within. Brian Foote and Joseph Yoder, "Big Ball of Mud," *Fourth Conference on Pattern Languages of Programs,* Monticello, IL, September 1997; in *Pattern Languages of Program Design 4,* ed. Brian Foote, Neil Harrison, and Hans Rohnert, chapter 29 (Boston: Addison-Wesley, 2000), available online: http://www.laputan.org/mud/.

45 **the software inside Toyota vehicles:** Koopman, "Case Study of Toyota Unintended Acceleration," slide 38. Koopman notes that spaghetti code can generate these high metrics of complexity.

Practices such as the MISRA Software Guidelines, issued by The Motor Industry Software Reliability Association in the UK, have been developed to better ensure that these systems are safe.

46 **the story of the Bayonne Bridge:** Philip K. Howard, *The Rule of Nobody: Saving America from Dead Laws and Broken Government* (New York: W. W. Norton, 2014).

46 **rules and regulations dictating the procedure:** Howard, *Rule of Nobody,* 8.

46 **public projects taking around ten years:** Howard, *Rule of Nobody,* 12. A project to replace another bridge took ten years to approve; the average duration of environmental review for highway projects is more than eight years.

46 **growth of rule systems as "regulatory accumulation":** Michael Mandel and Diana G. Carew, *Regulatory Improvement Commission: A Politically-Viable Approach to U.S. Regulatory Reform,* Progressive Policy Institute policy memo (Washington, DC: Progressive Policy Institute, May 2013), available online: http://www.progressivepolicy.org/wp-content/uploads/2013/05/05.2013-Mandel-Carew_Regulatory-Improvement-Commission_A-Politically-Viable-Approach-to-US-Regulatory-Reform.pdf.

47 **concept of *interoperability*:** John Palfrey and Urs Gasser, *Interop: The Promise and Perils of Highly Interconnected Systems* (New York: Basic Books, 2012).

47 **interdependence between different kinds of technologies:** Sergey V. Buldyrev et al., "Catastrophic Cascade of Failures in Interdependent Networks," *Nature* 464 (2010): 1025–28, http://

polymer.bu.edu/hes/articles/bppsh10.pdf. See also Natalie Wolchover, "Treading Softly in a Connected World," *Quanta Magazine,* March 18, 2013, https://www.quantamagazine.org/20130318-treading-softly-in-a-connected-world/.

48 **the cost of failure:** Cost of failure is really a hypothetical distribution, not a single number, but it could be thought of in terms of an expected value or average.

48 **Northeast Blackout in 2003:** J. R. Minkel, "The 2003 Northeast Blackout—Five Years Later," *Scientific American,* August 13, 2008, http://www.scientificamerican.com/article/2003-blackout-five-years-later/.

49 **For a long time, this approach worked:** These insights on the changes in cost of construction and cost of failure are thanks to Edward Jung (personal communication, March 28, 2014).

49 **Apple Maps mislabeled a supermarket:** "A Maps App with Problems," *The New York Times,* September 27, 2012, http://www.nytimes.com/slideshow/2012/09/27/technology/pogue-maps-ss-3.html.

49 **poliovirus has been reconstructed:** Jeronimo Cello et al., "Chemical Synthesis of Poliovirus cDNA: Generation of Infectious Virus in the Absence of Natural Template," *Science* 297 (2002): 1016–18; Eckard Wimmer, "The Test-Tube Synthesis of a Chemical Called Poliovirus: The Simple Synthesis of a Virus Has Far-Reaching Societal Implications," *EMBO Reports* 7, no. 1S (2006): S3–S9, http://www.ncbi.nlm.nih.gov/pmc/articles/PMC1490301/.

50 **a basic imperative of technology:** For example, see Doug Hill, *Not So Fast: Thinking Twice about Technology* (Cellarius Press, 2013, out of print; Athens, GA: University of Georgia Press, forthcoming); Kevin Kelly, *What Technology Wants.*

50 **Dijkstra noted that programming a computer:** Dijkstra, "On the Cruelty of Really Teaching Computing Science," E. W. Dijkstra Archive: The manuscripts of Edsger W. Dijkstra, 1930–2002, document no. EWD1036, December 1988, http://www.cs.utexas.edu/users/EWD/ewd10xx/EWD1036.PDF [hand-printed original], http://www.cs.utexas.edu/users/EWD/transcriptions/EWD10xx/EWD1036.html [typed transcript].

53 **Systems we build to reflect the world:** That the complexity of the world is reflected in the complexity of our systems is also discussed in Vikram Chandra, *Geek Sublime: The Beauty of Code, the Code of Beauty* (Minneapolis: Graywolf Press, 2014). One need not always end up with messy code because the world is messy, but it does often happen. Fortunately, there are ways to mitigate it. See Steve McConnell, *Code Complete: A Practical Handbook of Software Construction,* 2nd ed. (Redmond, WA: Microsoft Press, 2004), 583.

53 **building a self-driving vehicle:** The complexity of building self-driving cars was discussed by Google[x]'s "Captain of Moonshots" in his closing keynote address at South by Southwest Interactive (SXSW) 2015: Astro Teller, "How to Make Moonshots," *Backchannel,* March 17, 2015, https://medium.com/backchannel/how-to-make-moonshots-65845011a277.

53 **the exceptions that nonetheless have to be dealt with:** One solution is to use humans to manually troubleshoot, or at least hard-code, the exceptions. For example, here's how Google does this for Maps: "This is a Google-y approach to the problem of ultra-reliability. Many of Google's famously computation driven projects—like the creation of Google Maps—employed literally thousands of people to supervise and correct the automatic systems. It is one of Google's open secrets that they deploy human intelligence as a catalyst. Instead of programming in that last little bit of reliability, the final 1 or 0.1 or 0.01 percent, they can deploy a bit of cheap human brainpower. And over time, the humans work themselves out of jobs by teaching the machines how to act. 'When the human says, "Here's the right thing to do," that becomes something we can bake into the system and that will happen slightly less often in the future,' Teller said." Alexis C. Madrigal, "Inside Google's Secret Drone-Delivery Program," *The Atlantic,* August 28, 2014, http://www.theatlantic.com/technology/archive/2014/08/inside-googles-secret-drone-delivery-program/379306/?single_page=true.

55 **Scholars think it might have been an error:** "Friar Daw's Reply," from *Six Ecclesiastical Satires,* ed. James M. Dean, TEAMS Middle English Texts Series (Kalamazoo, MI: Medieval Institute Publications, 1991); available online at Robbins Library Digital Projects, University of Rochester, accessed April 30, 2015, http://d.lib.rochester.edu/teams/text/dean-six-ecclesiastical-satires-friar-daws-reply.

55 **more commonly, a *long tail:*** Note that not all heavy-tailed distributions, or long tails, are necessarily power laws.

56 **Often about half of the words:** András Kornai, *Mathematical Linguistics* (London: Springer-Verlag, 2008), 71. According to this source, the percentage of the words in a corpus that occur only once each—hapax legomena—is about 40–60 percent for many corpora.

56 **To avoid losing our exceptions and edge cases:** Related ideas are explored, along with the notion of language as a complex system, in William A. Kretzschmar Jr., *The Linguistics of Speech* (Cambridge, UK: Cambridge University Press, 2009). Also related is Firthian linguistics, "based on the view that language patterns cannot be accounted for in terms of a single system of analytic principles and categories," but that multiple different context-dependent systems may be called for. David Crystal, ed., *Dictionary of Linguistics and Phonetics,* 6th ed. (Malden, MA: Wiley-Blackwell, 2008), 181.

56 **Peter Norvig, Google's director of research:** Peter Norvig, "On Chomsky and the Two Cultures of Statistical Learning," accessed April 30, 2015, http://norvig.com/chomsky.html.

57 **great, though apocryphal, story:** There seem to be many versions of this apocryphal machine translation tale.

57 **What techniques are used by experts:** Nick Bostrom, *Superintelligence: Paths, Dangers, Strategies* (Oxford, UK: Oxford University Press, 2014), 15.

58 **say, 99.9 percent of the time:** I made these numbers up for effect, but if any linguist wants to chat, please reach out!

58–59 **"based on millions of specific features":** Alon Halevy et al., "The Unreasonable Effectiveness of Data," *IEEE Intelligent Systems* 24, no. 2 (2009): 8–12. In some ways, these statistical models are actually simpler than those that start from seemingly more elegant rules, because the latter end up being complicated by exceptions.

59 **sophisticated machine learning techniques:** See Douglas Heaven, "Higher State of Mind," *New Scientist* 219 (August 10, 2013), 32–35, available online (under the title "Not Like Us: Artificial Minds We Can't Understand"): http://complex.elte.hu/~csabai/simulationLab/AI_08_August_2013_New_Scientist.pdf.

59 **Frederick P. Brooks Jr. has noted:** Brooks, *Mythical Man-Month,* 183–84. Brooks recognizes many types of complexity, including that imposed by the environment the software must interact with.

60 **law turns out to look like a *fractal:*** David G. Post and Michael B. Eisen, "How Long Is the Coastline of the Law? Thoughts on the Fractal Nature of Legal Systems," *Journal of Legal Studies* 29, no. 2 J(2000): 545–84.

60 **legal scholar Jack Balkin discusses this:** Jack M. Balkin, "The Crystalline Structure of Legal Thought," *Rutgers Law Review* 39, no. 1 (1986): 1–108, http://www.yale.edu/lawweb/jbalkin/articles/crystal.pdf; Yale Law School Faculty Scholarship Series, Paper 294.

61 **The law professor David Post and the biologist Michael Eisen:** Post and Eisen, "How Long Is the Coastline of the Law?"

61 **they find features indicative of fractals:** Post and Eisen find power laws.

61 **"the value of good contracts and good lawyering":** Mark D. Flood and Oliver Goodenough, "Contract as Automaton: The Computational Representation of Financial Agreements," OFR (Office of Financial Research) Working Paper no. 15-04, March 26, 2015, https://financialresearch.gov/working-papers/files/OFRwp-2015-04_Contract-as-Automaton-The-Computational-Representation-of-Financial-Agreements.pdf.

62 **physics-trained sociologist Duncan Watts:** Duncan Watts, "Too Complex to Exist," *The Boston Globe,* June 14, 2009, http://www.boston.com/bostonglobe/ideas/articles/2009/06/14/too_complex_to_exist/?page=full.

62–63 **scholars have spoken of finding the optimal levels:** Palfrey and Gasser, *Interop.* See, for example, the chapter on complexity.

64 **if each module in a system:** Assuming that each distinct module can connect to another one in three distinct ways outbound and three distinct ways inbound, including self-loops, we have approximately 10^{32} potential networks. The number of stars in the observable universe is not known, but one estimate places it at fewer than 10^{30}. Elizabeth Howell, "How Many Stars Are In The Universe?" Space.com, May 31, 2014, http://www.space.com/26078-how-many-stars-are-there.html.

65 **good computer science and engineering practices:** While the main focus of this book is how to navigate through an

age of incomprehensibility that is already upon us, there are ways to design and construct more-manageable engineered systems, such as using "systems thinking." For example, see Nancy G. Leveson, *Engineering a Safer World: Systems Thinking Applied to Safety* (Cambridge, MA: The MIT Press, 2011).

65 **methods that can reduce the number of bugs:** McConnell, *Code Complete,* 521.

CHAPTER 3: LOSING THE BUBBLE

67 **In 1985, a patient entered a clinic:** Story and analysis from Nancy G. Leveson and Clark S. Turner, "An Investigation of the Therac-25 Accidents," *Computer* 26, no. 7 (1993), 18–41.

68 **"software does not degrade":** Quoted in Leveson and Turner, "An Investigation."

69 **the way machines count:** Machines—or more precisely, programming languages—can of course also enumerate starting from one, but many programming languages today count from zero. The reasons are old and have been forgotten by most programmers, but a good discussion of the history is Michael Hoye, "Citation Needed," *blarg? Mike Hoye's weblog,* October 22, 2013, http://exple.tive.org/blarg/2013/10/22/citation-needed/.

69 **the writer Scott Rosenberg notes:** Scott Rosenberg, *Dreaming in Code: Two Dozen Programmers, Three Years, 4,732 Bugs, and One Quest for Transcendent Software* (New York: Three Rivers Press, 2008), 6–7.

70 **"suddenly become opaque and bewildering":** Homer-Dixon, *The Ingenuity Gap,* 186.

72 **100 billion sentences:** Actually, to avoid duplicate sentences, it's really 10,000 nouns × 1,000 verbs × 9,999 nouns. It would still take more than 30,000 years to go through these sentences.

73 **from the linguist Steven Pinker:** Steven Pinker, *The Language Instinct: How the Mind Creates Language* (New York: William Morrow, 1994; repr. HarperPerennial, 1995), 205.

74 **"This is the cheese":** Quoted in Ray Kurzweil, *The Age of Spiritual Machines: When Computers Exceed Human Intelligence* (New York: Penguin, 1999), 95.

74 **Consider Kant Generator:** Program via Mark Pilgrim, *Dive into Python: Python from Novice to Pro,* updated 2004. Available free online: http://www.diveintopython.net/xml_processing/. While this program allows for embedding clauses within others, it seems that Kant Generator is not in fact infinitely recursive, as it will always terminate.

74 **structures known as *garden path sentences:*** This example is from "Garden Path Sentence," Wikipedia, accessed April 29, 2015, http:/en.wikipedia.org/wiki/Garden_path_sentence.

75 **our mental-processing limits:** Another example is the number of steps we think through when strategizing our decisions based on what we think others think and will do, as described by game theory. Few people think many levels deep ("I will do something based on what she thinks I think that she thinks that I think . . ."). For further reading, see Colin Camerer et al., "Behavioural Game Theory: Thinking, Learning and Teaching," Caltech Working Paper, http://people.hss.caltech.edu/~camerer/web_material/Ch08Pg_119-179.pdf.

75 **For most of us, it's about seven:** George A. Miller, "The Magical Number Seven, Plus or Minus Two: Some Limits on Our Capacity for Processing Information," *The Psychological Review* 63 (1956): 81–97, accessed April 30, 2015, http://www.musanim.com/miller1956/.

75 **eight seconds to transfer:** Lin Zhong, "Limitations of Human Mind," lecture notes for ELEC513/COMP513, Complexity in Modern Systems, Department of Electrical and Computer Engineering, Rice University, accessed April 30, 2015, http://www.ruf.rice.edu/~mobile/elec513/humanlimit-2.html.

76 **million times slower than a computer circuit:** Bostrom, *Superintelligence,* 59–60. For long-term memory, other estimates of storage capacity give vastly larger volumes, such as 2.5 petabytes. http://www.scientificamerican.com/article/what-is-the-memory-capacity/.

76 **delicately optimized by evolution:** Thomas Hills and Ralph Hertwig, "Why Aren't We Smarter Already: Evolutionary Trade-Offs and Cognitive Enhancements," *Current Directions in Psychological Science* 20, no. 6 (2011): 373–77.

76 **"Funes the Memorious":** Jorge Luis Borges, "Funes, His Memory," in *Collected Fictions,* trans. Andrew Hurley (New York: Viking Penguin, 1998), 131–37. I have chosen to retain an alternative, quite common translation of the story's title.

77 **the mathematician Srinivasa Ramanujan:** Robert Kanigel, *The Man Who Knew Infinity: A Life of the Genius Ramanujan* (New York: Charles Scribner's Sons, 1991; repr. Washington Square Press, 1992).

77 **programmer and novelist Vikram Chandra:** Chandra, *Geek Sublime,* 48.

78 **what is known as The Knowledge:** Jody Rosen, "The Knowledge, London's Legendary Taxi-Driver Test, Puts Up a Fight in the Age of GPS," *T, The New York Times Style Magazine,* November 10, 2014, http://tmagazine.blogs.nytimes.com/2014/11/10/london-taxi-test-knowledge/.

78 **outliers, impressive as they are, have limits:** César Hidalgo coined a term for the quantity of information and knowledge any one individual's head can hold: a personbyte. Hidalgo, *Why Information Grows.*

79–80 **John Symons and Jack Horner at the University of Kansas:** John Symons and Jack Horner, "Software Intensive Science," *Philosophy and Technology* 27, no. 3 (2014): 461–77.

81 **sometimes at the expense of human meaning:** Stanisław Lem examined this kind of incomprehensible complexity several decades ago in *Summa Technologiae,* trans. Joanna Zylinska (orig. pub. in Polish, 1964; Minneapolis: University of Minnesota Press, 2013), 96–97.

81 **Google recently turned powerful computational methods:** Jim Gao, "Machine Learning Applications for Data Center Optimization," Google White Paper, www.google.com/about/datacenters/efficiency/internal/assets/machine-learning-applicationsfor-datacenter-optimization-finalv2.pdf.

81 **To quote Google's blog:** Joe Kava, "Better Data Centers through Machine Learning," *Google Official Blog,* May 28,

2014, http://googleblog.blogspot.com/2014/05/better-data-centers-through-machine.html.

82 **it wouldn't necessarily be meaningful:** Douglas Heaven, "Higher State of Mind," *New Scientist* 219 (August 10, 2013), 32–35, available online (under title "Not Like Us: Artificial Minds We Can't Understand"): http://complex.elte.hu/~csabai/simulationLab/AI_08_August_2013_New_Scientist.pdf.

83 **design a type of computer circuit:** Note that this circuit was actually evolved in hardware (each member of the population was instantiated in a field-programmable gate array—a type of programmable circuit—rather than being simulated), while many evolutionary algorithms occur entirely in software. Adrian Thompson, "Exploring Beyond the Scope of Human Design: Automatic Generation of FPGA Configurations Through Artificial Evolution (Extended Abstract)," 8th Annual Advanced PLD and FPGA Conference, 1998, https://web.archive.org/web/20101215100211/http://www.informatics.sussex.ac.uk/users/adrianth/ascot/paper.ps. See also Thompson, "An Evolved Circuit, Intrinsic in Silicon, Entwined with Physics," in *Evolvable Systems: From Biology to Hardware,* ed. Tetsuya Higuchi et al., Lecture Notes in Computer Science vol. 1259 (New York: Springer, 2008), 390–405, available online: http://citeseerx.ist.psu.edu/viewdoc/download?doi=10.1.1.50.9691&rep=rep1&type=pdf.

83 **"Not only is it ugly":** Kevin Kelly, *Out of Control* (New York: Basic Books, 1994), 338.

84 **In the realm of logistics:** Steven Rosenbush and Laura Stevens, "At UPS, the Algorithm Is the Driver," *The Wall Street Journal,* February 16, 2015, http://www.wsj.com/articles/at-ups-the-Algorithm-is-the-driver-1424136536. On the blog *Marginal Revolution,* Alex Tabarrok refers to this kind of intelligence as "opaque intelligence." http://marginalrevolution.com/marginalrevolution/2015/02/opaque-intelligence.html.

84 **the economist Tyler Cowen noted:** Tyler Cowen, *Average Is Over: Powering America beyond the Age of the Great Stagnation* (New York: Dutton, 2013), 72.

84 **"both praised and panned":** Feng-Hsiung Hsu, *Behind Deep Blue: Building the Computer That Defeated the World Chess Champion* (Princeton, NJ: Princeton University Press, 2002).

85 **"slow, tortuous reading":** Flood and Goodenough, "Contract as Automaton."

85 **when forty-five tax professionals:** Donald Sull and Kathleen M. Eisenhardt, *Simple Rules: How to Thrive in a Complex World* (New York: Houghton Mifflin Harcourt, 2015), 12–13.

86 **I have on my shelf three books:** Paula Findlen, ed., *Athanasius Kircher: The Last Man Who Knew Everything* (New York: Routledge, 2004); Andrew Robinson, *The Last Man Who Knew Everything: Thomas Young, the Anonymous Polymath Who Proved Newton Wrong, Explained How We See, Cured the Sick and Deciphered the Rosetta Stone* (New York: Plume, 2007; orig. pub. Saddle River, NJ: Pi Press / Pearson Education, 2006); Leonard Warren, *Joseph Leidy: The Last Man Who Knew Everything* (New Haven: Yale University Press, 1998).

87 **As the writer Philip Ball notes:** Philip Ball, *Curiosity: How Science Became Interested in Everything* (Chicago: University of Chicago Press, 2013), 55.

87–88 **point made by the writer Patrick Mauries:** Ball, *Curiosity,* 156.

89 **According to the scholar Daniel Boorstin:** Daniel J. Boorstin, *The Discoverers: A History of Man's Search to Know His World and Himself* (New York: Random House, 1983), 414.

89 **the words of Frederick the Great:** Quoted in Boorstin, *The Discoverers,* 414.

89 **some professors chose their titles:** Ball, *Curiosity,* 120.

89 **the mathematician Isaac Barrow noted:** Quoted in Ball, *Curiosity,* 120.

90 **"more and more about less and less":** John M. Ziman, *Knowing Everything About Nothing: Specialization and Change in Research Careers* (Cambridge, UK: Cambridge University Press, 1987), v.

90 **a theory about the "burden of knowledge":** Benjamin F. Jones, "The Burden of Knowledge and the 'Death of the Renaissance Man': Is Innovation Getting Harder?" *The Review of Economic Studies* 76 (2009): 283–317. This theory is primarily focused on knowledge related to technological progress.

90 **In one article coauthored by Jones:** Benjamin F. Jones, E. J. Reedy, and Bruce A. Weinberg, "Age and Scientific Genius," in *The Wiley Handbook of Genius,* ed. Dean Keith Simonton (Malden, MA: John Wiley and Sons, 2014).

90 **more than 36 million books:** "Fascinating Facts," Library of Congress, accessed March 2, 2015, http://www.loc.gov/about/fascinating-facts/.

90 **The biologist E. O. Wilson described the change:** Edward O. Wilson, *Consilience: The Unity of Knowledge* (New York: Alfred A. Knopf, 1998; repr. Vintage Books, 1999), 42–43.

91 **driverless cars is a good example:** Jordan Bell-Masterson, "Innovation Series: The Rising Costs of Invention," *Growthology,* March 24, 2015, http://www.kauffman.org/blogs/growthology/2015/03/innovation-series-increasing-costs-of-invention.

92 **attempt to visualize these patterns of teamwork:** Michael Ogawa and Kwan-Liu Ma, "Software Evolution Storylines," *SOFTVIS '10: Proceedings of the 5th International Symposium on Software Visualization* (New York: ACM Digital Library, 2010), 35–42, available online: http://vis.cs.ucdavis.edu/papers/softvis_storylines.pdf. Also see http://www.michaelogawa.com/.

93 **"Of all the monsters who fill the nightmares":** Brooks, *Mythical Man-Month,* 180.

CHAPTER 4: OUR BUG-RIDDEN WORLD

95 **the video game Galaga:** "But the questions always nagged for the past decades. Was the cheat purposefully added to the code as a backdoor for players in-the-know? Or is the cheat a glitch in the software—an unexpected side effect that persisted over several releases of the game ROMs? If the cheat is a glitch,

what is wrong with the code? Is there another sequence of actions to get to the disabled state faster?" Chris Cantrell, "Arcade / Galaga," Computer Archeology, accessed April 29, 2015, http://computerarcheology.com/Arcade/Galaga/. I have also seen an argument in favor of its being an intentional cheat, as discussed here: Jason Eckert, "The Galaga No Fire Cheat Mystery," October 31, 2012 (updated May 2014), http://triosdevelopers.com/jason.eckert/blog/Entries/2012/10/31_The_Galaga_no_fire_cheat_mystery.html. I first learned of the Galaga glitch from Clive Thompson, *Smarter Than You Think: How Technology Is Changing Our Minds for the Better* (New York: Penguin, 2013).

96 **how we realize that we are in the Entanglement:** Distinguished Google Fellow Urs Hölzle: "Complexity is evil in the grand scheme of things because it makes it possible for these bugs to lurk that you see only once every two or three years, but when you see them it's a big story because it had a large, cascading effect." Jack Clark, "Google: 'At Scale, Everything Breaks,'" *ZDNet,* June 22, 2011, http://www.zdnet.com/article/google-at-scale-everything-breaks/2/.

96 **In 1950, Alan Turing noted:** A. M. Turing, "Computing Machinery and Intelligence," *Mind* 59 (1950): 433–60. Widely available online, e.g., http://cogprints.org/499/1/turing.html.

97 **a widely used simulator of gravitation:** There are about 10,000 mixed-precision instances (the specific type of error) across about 1,000 lines out of approximately 30,000 lines of code.

John Symons and Jack Horner, "Software Intensive Science," *Philosophy and Technology* 27, no. 3 (2014): 461–77; J. K. Horner, "Persistence of Plummer-Distributed Small Globular Clusters as a Function of Primordial-Binary Population Size," *Proceedings of the International Conference on Scientific Computing (CSC)* (Athens: CSREA Press, 2013), 100–106.

97 **These bugs range:** One estimate places the number of bugs at one every three to five lines of code, an astonishingly high number. Roger A. Grimes, "In His Own Words: Confessions of a Cyber Warrior," *InfoWorld,* July 9, 2013, http://www.infoworld.com/article/2611471/security/in-his-own-words—confessions-of-a-cyber-warrior.html?page=3.

98 **In 1996, a computer "bug detective" published:** Bruce Brown, Nigel R. Smith, and Bruce Kratofil, *The Windows 95 Bug Collection* (Reading, MA: Addison-Wesley, 1996), 3–10.

98 **when a software project becomes twice as large:** McConnell, *Code Complete,* 652.

99 **Take the Boeing 777:** J. M. Carlson and John Doyle, "Complexity and Robustness." *PNAS* 99, Suppl. 1 (2002): 2538–45.

99 **According to two scientists:** Carlson and Doyle, "Complexity and Robustness."

100 **the computer scientist Philip Koopman has noted:** Koopman, "Case Study of Toyota Unintended Acceleration," slide 20.

100 **has released a series of books:** See the fascinating works by Kate Ascher: *The Works: Anatomy of a City* (2005); *The Heights: Anatomy of a Skyscraper* (2011); and *The Way to Go: Moving by Sea, Land, and Air* (2015), all from The Penguin Press, New York.

101 **a water main broke in Weston:** For further reading, see "Massachusetts Water Crisis," *The Boston Globe,* http://www.boston.com/news/local/massachusetts/specials/Water_crisis/.

101 **the physical infrastructure of the Internet:** Andrew Blum, *Tubes: A Journey to the Center of the Internet* (New York: Ecco, 2012).

102 **All the bugs cannot be eradicated:** This point is also discussed in Roger A. Grimes, "Shellshock Proves Open Source's 'Many Eyes' Can't See Straight," *InfoWorld,* September 30, 2014, http://www.infoworld.com/article/2689233/security/shellshock-proves-open-source-many-eyes-wrong.html.

103 **Gmail—Google's email service—suffered an outage:** Jon Brodkin, "Why Gmail Went Down: Google Misconfigured Load Balancing Servers (Updated)," *Ars Technica,* December 11, 2012, http://arstechnica.com/information-technology/2012/12/why-gmail-went-down-google-misconfigured-chromes-sync-server/.

104 **a blog post from Google in 2006:** Joshua Bloch, "Extra, Extra—Read All About It: Nearly All Binary Searches and Mergesorts Are Broken," *Google Research Blog,* June 2, 2006, http://googleresearch.blogspot.com/2006/06/extra-extra-read-all-about-it-nearly.html. Also discussed in Chandra, *Geek Sublime,* 124.

105 **The bug is a window:** A glitch is "a possibility to glance at software's inner structure." Olga Goriunova and Alexei Shulgin, "Glitch," in *Software Studies: A Lexicon,* ed. Matthew Fuller, 110–19 (Cambridge, MA: The MIT Press, 2008), 114. Errors and bugs as a window into improving a system is a concept

also explored by Nassim Nicholas Taleb in *Antifragile: Things That Gain from Disorder* (New York: Random House, 2012). For an application of Taleb's concept of antifragility to software development, see Martin Monperrus, "Principles of Antifragile Software," http://arxiv.org/pdf/1404.3056.pdf.

105 **at the beginning of 1982, the Vancouver Stock Exchange:** This story about the Vancouver Stock Exchange index has been told in numerous places. I have primarily drawn from these sources: Anne Greenbaum and Timothy P. Chartier, *Numerical Methods: Design, Analysis, and Computer Implementation of Algorithms* (Princeton, NJ: Princeton University Press, 2012), 117; Kevin Quinn, "Ever Had Problems Rounding Off Figures? This Stock Exchange Has," *Wall Street Journal,* November 8, 1983.

107 **software known as Chaos Monkey:** Chaos Monkey is available online at https://github.com/Netflix/SimianArmy/wiki/Chaos-Monkey.

109 **The physicist Enrico Fermi:** Quoted in Bill Bryson, *A Short History of Nearly Everything* (New York: Broadway Books, 2003), 162.

110 **only 53 kilobytes of high-speed RAM:** Data point found in George Dyson, *Turing's Cathedral: The Origins of the Digital Universe* (New York: Pantheon, 2012), 4.

CHAPTER 5: THE NEED FOR BIOLOGICAL THINKING

111 **English physician named Nathanael Fairfax:** Brief biographical data is from the *Dictionary of National Biography, 1885–*

1900, vol. 18, "Fairfax, Nathaniel, M.D.," http://en.wikisource .org/wiki/Page:Dictionary_of_National_Biography_volume_18.djvu/143. Note that the given name is spelled differently on his published articles, but the biographical details make clear that this is the same person.

111 **"Divers Instances of Peculiarities of Nature":** Nathanael Fairfax, "Divers Instances of Peculiarities of Nature, Both in Men and Brutes; Communicated by the Same," *Philosophical Transactions* 1666–67, 2 (1666): 549–51.

112 **back in Woolsthorpe:** Newton also spent some time at Cambridge during these plague years, where he did work as well. Information on Newton's life and work can be found in these sources: George Smith, "Isaac Newton," *The Stanford Encyclopedia of Philosophy,* Fall 2008 edition, ed. Edward N. Zalta, http://plato .stanford.edu/archives/fall2008/entries/newton/; V. Frederick Rickey, "Isaac Newton: Man, Myth, and Mathematics," *The College Mathematics Journal* 18, no. 5 (1987): 362–89.

113 **"great tragedy of Science":** Tania Lombrozo, "Must Science Murder Its Darlings?" *NPR 13.7: Cosmos and Culture,* January 27, 2014, http://www.npr.org/blogs/13.7/2014/01/26/266784786/must -science-murder-its-darlings.

114 **The physicist Freeman Dyson has described:** Freeman J. Dyson, *Infinite in All Directions,* repr. ed. (New York: Harper Perennial, 2004; orig. pub. 1988), 40.

115 **biologists, as a rule, have a greater comfort:** While Darwin was a unifying force, he was also clearly a diversifier, with great attention to complex detail. Shane Parrish, "What Made

Charles Darwin an Effective Thinker? Follow the Golden Rule," *Farnam Street,* January 11, 2016, https://www.farnamstreetblog.com/2016/01/charles-darwin-thinker/. Other fields outside science have a "biological" tendency as well. For example, historians tend away from abstraction and generalization in order to understand history.

116 **"first, you assume a spherical cow":** Biologists are very much against the idea of the spherical cow: "If biologists are much like physicists in stretching the limits of experimental reductionism, they are also like engineers in revelling in the enormity, variety and sheer complexity of the systems they study. No interest in spherical cows here." John Doyle, "Computational Biology: Beyond the Spherical Cow," *Nature* 411 (2001): 151–52, http://www.nature.com/nature/journal/v411/n6834/full/411151a0.html.

119 **the biologist Steven Benner notes:** Steven A. Benner, "Aesthetics in Synthesis and Synthetic Biology," *Current Opinion in Chemical Biology* 16, no. 5–6 (2012): 581–85.

119 **obsolete legacy code, just as technology:** Obsolete legacy genetic code is similar to the "cruft" found in software: extra material that is no longer necessary for the current version's function, yet can last far longer than we might wish.

119 **a number of honey locust trees:** Whit Bronaugh, "The Trees That Miss the Mammoths," *American Forests,* Winter 2010, http://www.americanforests.org/magazine/article/trees-that-miss-the-mammoths/.

120 **Biology handles legacy code differently:** Obsolete legacy software code may not be pruned away by natural selection, but it can be removed via periodic purges and cleaning of the code, making it a bit more similar to biology.

123 **field biologists for technology:** The term "digital biologist" is used in a somewhat similar manner by the historian George Dyson in "A Universe of Self-Replicating Code," *Edge,* accessed February 5, 2015, http://edge.org/conversation/a-universe-of-self-replicating-code.

123 **initial steps in the discovery:** "RNAi," *NOVA,* July 26, 2005, http://www.pbs.org/wgbh/nova/body/rnai.html.

124 **Isaac Asimov is reputed to have noted:** Howard Wainer and Shaun Lysen, "That's Funny," *American Scientist* 97, no. 4 (2009): 272, http://www.americanscientist.org/issues/pub/thats-funny.

125 **one way that new drugs are created:** Dan Hurley, "Why Are So Few Blockbuster Drugs Invented Today?" *The New York Times Magazine,* November 13, 2014, http://www.nytimes.com/2014/11/16/magazine/why-are-there-so-few-new-drugs-invented-today.html. This point about what we can learn from testing pharmaceuticals was made to me by Edward Jung.

126 **Stewart Brand noted about legacy systems:** Stewart Brand, *The Clock of the Long Now: Time and Responsibility* (New York: Basic Books, 1999), 85.

126 **a partial meltdown at the Three Mile Island plant:** Peter G. Neumann, *Computer-Related Risks* (New York: ACM Press, 1995), 122.

128 **elaborates on the structure of the pantheon:** Neal Stephenson, *Cryptonomicon* (New York: Avon Books, 1999; repr. 2002), 802–3.

130 **Corky Ramirez:** Note that in the episode "The Van Buren Boys," someone is referred to as "Ramirez" in a bar (though I believe his name is stressed differently than Kramer's pronunciation of Corky Ramirez). Perhaps he is visible in the room, but it is unclear. *Seinfeld* superfans: please send me mail.

130 **delightfully evocative term: "greeblies":** Or, alternatively, "greebles." Kelly, *What Technology Wants,* 318.

130 **the mathematician Benoit Mandelbrot:** Benoit B. Mandelbrot, *The Fractal Geometry of Nature* (New York: W. H. Freeman and Company, 1982), 1.

131 **Recall "Funes the Memorious":** Borges, "Funes, His Memory," in *Collected Fictions,* 131–37.

136 **"The patterns of a river network":** Philip Ball, *Branches,* vol. 3 of *Nature's Patterns: A Tapestry in Three Parts* (Oxford, UK: Oxford University Press, 2009), 181.

136 **researchers analyzed the United States Code:** William Li et al., "Law Is Code: A Software Engineering Approach to Analyzing the United States Code," *Journal of Business and Technology Law* 10, no. 2 (2015): 297–372.

138 **The ideas of these philosophers:** Jonathan Barnes, ed. and trans., *Early Greek Philosophy* (London: Penguin Classics, 1987). Also see *The Stanford Encyclopedia of Philosophy,* http://plato.stanford.edu/.

139 **"evaporated by the sun":** Barnes, *Early Greek Philosophy,* 72.

139 **the idea of the Greek *kosmos*:** The discussion of the nature of *kosmos* and *arche* is from Barnes, *Early Greek Philosophy.*

140 **According to Philip Ball, anomalies and eccentricities:** Ball, *Curiosity,* 98–99.

140 **a strong penchant in the early days of science:** Lorraine Daston, "The Language of Strange Facts in Early Modern Science," in *Inscribing Science: Scientific Texts and the Materiality of Communication,* ed. Timothy Lenoir (Stanford, CA: Stanford University Press, 1998), 20–38.

140 **"The first scientific facts":** Daston, "Language of Strange Facts," 38.

141 **examined hundreds of millions of interactions:** Johan Bollen et al., "Clickstream Data Yields High-Resolution Maps of Science," *PLoS ONE* 4, no. 3 (2009): e4803.

143 **first originated in computing education:** Nicholas Donofrio, Jim Spohrer, and Hossein S. Zadeh, "Research-Driven Medical Education and Practice: A Case for T-shaped Professionals," *MJA Viewpoint,* 2009, http://www.ceri.msu.edu/wp-content/uploads/2010/06/A-Case-for-T-Shaped-Professionals-20090907-Hossein.pdf.

144 **difficult to educate T-shaped people:** On the growth of specialization, Thomas Homer-Dixon has written of a "narrowing of expertise." Homer-Dixon, *The Ingenuity Gap,* 176.

144 **As business professor David Teece has noted:** David J. Teece, "A Dynamic Capabilities Perspective on Building Firm-Level

Competitiveness," slide 43, Tusher Center on Intellectual Capital, http://faculty.haas.berkeley.edu/lyons/teecetusherslides.pdf.

145 **the title character of *Hild*:** Nicola Griffith, *Hild: A Novel* (New York: Farrar, Straus and Giroux, 2013).

147 **science bookseller John Ptak notes:** John Ptak, "A Cloud Map (1873)," JF Ptak Science Books, January 13, 2015, http://longstreet.typepad.com/thesciencebookstore/2015/01/a-cloud-map-1873.html.

CHAPTER 6: WALKING HUMBLY WITH TECHNOLOGY

152 **"for man's intellect indubitably has a limit":** Moses Maimonides, *The Guide of the Perplexed,* vol. 1, trans. Shlomo Pines (Chicago: The University of Chicago Press, 1963), 65–66.

153 **"queerer than we *can* suppose":** J. B. S. Haldane, *Possible Worlds and Other Essays* (London: Chatto & Windus, 1928), 286.

153 **limitations to what we can know:** For a further discussion on scientific humility, see Marcelo Gleiser, *The Island of Knowledge: The Limits of Science and the Search for Meaning* (New York: Basic Books, 2014).

154 **video game designer and writer Ian Bogost:** Ian Bogost, "The Cathedral of Computation," *The Atlantic,* January 15, 2015, http://www.theatlantic.com/technology/archive/2015/01/the-cathedral-of-computation/384300/.

154 **a perfect and immaculate process:** This is discussed further in Bogost, "Cathedral of Computation."

155 **the "humble programmer":** Edsger Dijkstra, "The Humble Programmer." *Communications of the ACM* 15, no. 10 (1972): 859–66.

155 **"Wisdom starts with epistemological modesty":** David Brooks, *The Road to Character* (New York: Random House, 2015), 263.

156 **nevertheless see a "glorious mess":** Carl Zimmer, "Is Most of Our DNA Garbage?" *The New York Times Magazine,* March 5, 2015, http://www.nytimes.com/2015/03/08/magazine/is-most-of-our-dna-garbage.html.

157 **The book includes maxims:** These examples are all from Appendix I of John Gall, *The Systems Bible: The Beginner's Guide to Systems Large and Small,* 3rd ed. (Walker, MN: The General Systemantics Press, 2003).

157 **a number of points similar to those:** Gall even admonishes his readers to "cherish your system-failures," just as biologists collect interesting observations (Gall cites Charles Darwin's example approvingly), and he headlines the admission that "large complex systems are beyond human capacity to evaluate." *Systems Bible,* xx, 68.

158 **humility in the face of systems:** See also Robert Herritt, "When Technology Ceases to Amaze," *The New Atlantis* 41, Winter 2014, 121–31, http://www.thenewatlantis.com/publications/when-technology-ceases-to-amaze.

158 **the designer Don Norman:** Donald A. Norman, *Living with Complexity* (Cambridge, MA: The MIT Press, 2010), 117–18.

159 **"percent-done progress indicators":** Daniel Engber, "Who Made That Progress Bar?" *The New York Times Magazine,* March 7, 2014, http://www.nytimes.com/2014/03/09/magazine/who-made-that-progress-bar.html.

159 **divorced from the underlying process:** Kate Greene, "How Should We Program Computers to Deceive?" *Pacific Standard,* September 3, 2014, http://www.psmag.com/nature-and-technology/technology-deception-elevator-crosswalk-programming-robots-lie-89669.

160 **friendly user interface of TurboTax:** Philip Guo tweeted: "i wonder how many layers of nested if-statements are in the code for TurboTax," March 26, 2014, https://twitter.com/pgbovine/status/448988621778194432. I thank Dan Katz for the insight of TurboTax as an interface for the law.

163 **tools that, in Gingold's words:** Chaim Gingold, *Miniature Gardens and Magic Crayons: Games, Spaces, and Worlds,* master's thesis, Georgia Institute of Technology, 2003, 62, http://levitylab.com/cog/writing/Games-Spaces-Worlds.pdf.

164 **systems are so completely automated:** For further reading, see Nicholas Carr, *The Glass Cage: How Our Computers Are Changing Us* (New York: W. W. Norton, 2014).

164 **"concealed electronic complexity":** Winner, *Autonomous Technology,* 285.

164 **component of the telephone system:** Eytan Adar et al., "Benevolent Deception in Human Computer Interaction," *CHI '13: Proceedings of the SIGCHI Conference on Human Factors in Computing Systems,* Paris, France, April 27–May 2, 2013 (New York: ACM Digital Library, 2013): 1863–72.

166 **the computer game SimCity:** For more on SimCity and how it can shed some light on how a complicated system works, read Doug Bierend, "SimCity That I Used to Know: On the Game's

25th Birthday, a Devotee Talks with Creator Will Wright," re:form, October 17, 2014, https://medium.com/re-form/sim city-that-i-used-to-know-d5d8c49e3e1d.

166 **Near the end of *Average Is Over:*** Cowen, *Average Is Over,* 227–28. Cowen is speculating specifically about the future of economics and other social sciences, but we could need such "interpreters" for our future understanding of anthropic systems as well.

167 **a glimmer of intuition into complex systems:** These interpreters will likely work in conjunction with machines, as Cowen suggests. We might become like "centaurs," but half machine instead of half horse. Clive Thompson, *Smarter Than You Think: How Technology Is Changing Our Minds for the Better* (New York: Penguin, 2013).

168 **David Cope of the University of California, Santa Cruz:** David Cope, "Experiments in Musical Intelligence," accessed April 30, 2015, http://artsites.ucsc.edu/faculty/cope/experiments.htm. See also Ryan Blitstein, "Triumph of the Cyborg Composer," *Pacific Standard,* February 22, 2010, http://www.psmag.com/books-and-culture/triumph-of-the-cyborg-composer-8507.

169 **His computational creations can provide him:** Cope even writes the following on his website, extending pride of parentage to all humanity: "The music our algorithms compose are just as much ours as the music created by the greatest of our personal human inspirations." Cope, "Experiments." And he is quoted as saying the following: "All the computer is is just an extension of me. They're nothing but wonderfully organized

shovels. I wouldn't give credit to the shovel for digging the hole. Would you?" Blitstein, "Triumph of the Cyborg Composer."

169 ***naches* is also a framework:** The roboticist Hans Moravec has referred to our more powerful descendants as "mind children," and a similar approach characterizes a short story by the science fiction writer Ted Chiang, in which technologically enhanced humans have long surpassed "regular" humans in their ability to make scientific discoveries. In the end, little to nothing is understood by (nonenhanced) humanity. But that's okay, because "We need not be intimidated by the accomplishments of metahuman science. We should always remember that the technologies that made metahumans possible were originally developed by humans, and they were no smarter than we." See Luke Muehlhauser and Nick Bostrom, "Why We Need Friendly AI," *Think* 36, no. 13 (Spring 2014), 41–47; and Ted Chiang, *Stories of Your Life and Others* (New York: Tor Books, 2003), 203.

170 **understand the most complex parts of the world:** In many cases, we might even want to have a technology too complex to understand, because it means that it is sophisticated and powerful.

170 **a grab bag of intriguing ideas:** *The World of Wonders: A Record of Things Wonderful in Nature, Science, and Art* (London: Cassell, Petter, and Galpin, exact year of publication unknown), https://archive.org/details/worldofwondersre00londrich.

171 **We tell ourselves simplifying stories:** This kind of simplifying storytelling is discussed by Philip Ball in "The Story Trap,"

Aeon, November 12, 2015, https://aeon.co/essays/why-story-is-used-to-explain-symphonies-and-sport-matches-alike.

172 **Don Norman has written of the delight:** Norman, *Living with Complexity.* 15.

172 **the definition of an infield fly:** MLB.com, "Official Rules: 2.00 Definition of Terms," accessed February 24, 2015, http://mlb.mlb.com/mlb/official_info/official_rules/definition_terms_2.jsp.

172 **to conflate mystery and wonder:** This is my own personal distinction between these two terms. No doubt there are many others.

173 **"sad inertness of a world":** Hubert Dreyfus and Sean Dorrance Kelly, *All Things Shining: Reading the Western Classics to Find Meaning in a Secular Age* (New York: Free Press, 2011), 88.

173 **we strive to eliminate our ignorance:** For a discussion of managing our ignorance of our technology, see Herritt, "When Technology Ceases to Amaze."

174 **shifting the car into neutral:** "Customer FAQs Regarding the Sticking Accelerator Pedal and Floor Mat Pedal Entrapment Recalls," Toyota Pressroom, accessed April 27, 2015, http://pressroom.toyota.com/article_print.cfm?article_id=1861.

174–175 **isn't the worst thing to tell someone:** I am thankful for this insight, as well the insights related to limitative theorems, from discussion with folks from the Department of Philosophy at the University of Kansas.

175 **incomprehensible systems are the new reality:** For example, just because we might not fully grasp all the details of a self-driving car, that doesn't mean that it can't be much safer than one driven by a person. And by the way, we already don't really understand the car driven by a person, let alone the driver himself!

176 **the "unthinkable present":** Quoted in Carlin Romano, *America the Philosophical* (New York: Alfred A. Knopf, 2012), 501.

Index

Page numbers beginning with 187 refer to notes.

INDEX

INDEX